THE BIRDS OF BRITAIN
THEIR DISTRIBUTION AND HABITS

THE BIRDS OF BRITAIN
THEIR DISTRIBUTION AND HABITS

BY

A. H. EVANS, M.A., F.Z.S., M.B.O.U.

Cambridge:
at the University Press
1916

CAMBRIDGE
UNIVERSITY PRESS

University Printing House, Cambridge CB2 8BS, United Kingdom

Cambridge University Press is part of the University of Cambridge.

It furthers the University's mission by disseminating knowledge in the pursuit of education, learning and research at the highest international levels of excellence.

www.cambridge.org
Information on this title: www.cambridge.org/9781107536876

© Cambridge University Press 1916

This publication is in copyright. Subject to statutory exception and to the provisions of relevant collective licensing agreements, no reproduction of any part may take place without the written permission of Cambridge University Press.

First published 1916
First paperback edition 2015

A catalogue record for this publication is available from the British Library

ISBN 978-1-107-53687-6 Paperback

Cambridge University Press has no responsibility for the persistence or accuracy of URLs for external or third-party internet websites referred to in this publication, and does not guarantee that any content on such websites is, or will remain, accurate or appropriate.

PREFACE

IT is hoped that this little work, though primarily intended for schools, may be found useful by those who require a short hand-book which includes the results of the most recent observations, and is adapted to modern nomenclature. The author wishes to tender his thanks to Mr W. Eagle Clarke, who is now preparing a new edition of Yarrell's *British Birds*, for glancing over the proofs, in order that they may be in general agreement with the forthcoming work; while he cannot fail to recall with great regret the times when he was working in company with the late Howard Saunders, then engaged in compiling his well-known *Manual of British Birds*, or short form of Yarrell.

A. H. E.

9 HARVEY ROAD, CAMBRIDGE,
24 *February*, 1916

CONTENTS

			PAGE
INTRODUCTION			1
Order I.	PASSERES		18
Order II.	PICARIÆ		98
Order III.	STRIGES		111
Order IV.	ACCIPITRES		116
Order V.	STEGANOPODES . . .		132
Order VI.	HERODIONES . . .		136
Order VII.	ODONTOGLOSSÆ . .		16, 260
Order VIII.	ANSERES		143
Order IX.	COLUMBÆ		163
Order X.	PTEROCLETES . . .		167
Order XI.	GALLINÆ		169
Order XII.	GRALLÆ		179
Order XIII.	LIMICOLÆ		188
Order XIV.	GAVIÆ		219
Order XV.	ALCÆ		236
Order XVI.	PYGOPODES		243
Order XVII.	TUBINARES		250
LIST OF OCCASIONAL VISITORS TO BRITAIN .			257
INDEX			263

ILLUSTRATIONS

			PAGE
1.	Archæopteryx lithographica. From *The Cambridge Natural History*, Vol. IX		2
2.	A Falcon. To shew the nomenclature of the external parts. From *The Cambridge Natural History*, Vol. IX		4
3.	The Zoo-Geographical Regions		14
4.	Ring-ousel	H. W. Richmond	21
5.	Wheatear	J. Holmes	23
6.	Redstart	K. J. A. Davis	25
7.	Robin	H. W. Richmond	27
8.	Nightingale at nest	,,	28
9.	Willow-Wren at nest	T. L. Smith	32
10.	Reed-Warbler's nest	J. Holmes	35
11.	Grasshopper-Warbler at nest	W. Farren	37
12.	Dipper	H. W. Richmond	40
13.	Dipper's nest	J. Holmes	41
14.	Bearded Tit at nest	W. Farren	42
15.	Long-tailed Tit	J. Holmes	44
16.	Long-tailed Tit's nest	,,	45
17.	Coal Tit	T. L. Smith	47
18.	Wren	H. W. Richmond	52
19.	Wren's nest	T. L. Smith	53
20.	Tree-Creeper	H. W. Richmond	54
21.	Pied Wagtail	J. Holmes	56
22.	Red-backed Shrikes	H. W. Richmond	61
23.	Waxwing in Zoological Gardens	,,	64

Illustrations

			PAGE
24.	Spotted Flycatcher	H. W. Richmond	65
25.	Swallow's nest	,,	67
26.	Sand-martins' burrows	,,	68
27.	Bullfinch	J. Holmes	78
28.	Corn-Bunting	H. W. Richmond	81
29.	Starling	J. Holmes	85
30.	Jay	,,	88
31.	Magpie	,,	90
32.	Magpie's nest	K. J. A. Davis	91
33.	Raven's nest	T. L. Smith	92
34.	Rook	J. Holmes	95
35.	Skylark	,,	96
36.	Swift on nest	H. W. Richmond	100
37.	Goatsucker and eggs	,,	101
38.	Woodpecker's nesting-holes	A. C. V. Gosset	104
39.	Green Woodpecker	J. Holmes	106
40.	Kingfisher	,,	107
41.	Hoopoes	H. W. Richmond	109
42.	Cuckoo	J. Holmes	110
43.	Tawny Owl	,,	114
44.	Griffon Vulture at Zoological Gardens	H. W. Richmond	116
45.	Marsh-Harrier	J. Holmes	118
46.	Golden Eagle at Zoological Gardens	H. W. Richmond	122
47.	Greenland Falcon at Zoological Gardens	,,	126
48.	Peregrine Falcon	J. Holmes	127
49.	Merlin's nest and eggs	T. L. Smith	129
50.	Kestrel	J. Holmes	131
51.	Cormorants and nests, on Farne Islands	H. W. Richmond	133
52.	Gannets nesting on Bass Rock	,,	135
53.	Common Heron	J. Holmes	137
54.	White Storks at Zoological Gardens	H. W. Richmond	141

Illustrations

			PAGE
55.	Spoonbill and Ibises at Zoological Gardens	H. W. Richmond	143
56.	Mute Swan on nest	T. L. Smith	145
57.	Sheldrake and nest	J. Holmes	150
58.	Nest and eggs of Teal	T. L. Smith	154
59.	Red-breasted Merganser	H. W. Richmond	162
60.	Rock-Dove	,,	166
61.	Hen Capercaillie	,,	170
62.	Red Grouse	J. Holmes	172
63.	Pheasant	,,	175
64.	Red-legged Partridge	,,	177
65.	Water-rail on nest	K. J. A. Davis	180
66.	Coot	J. Holmes	182
67.	Coot's nest and eggs	T. L. Smith	183
68.	Common Crane at Zoological Gardens	H. W. Richmond	185
69.	Great Bustards at Zoological Gardens	,,	186
70.	Stone-Curlew and eggs	,,	189
71.	Ringed Plover on nest	K. J. A. Davis	191
72.	Lapwing	J. Holmes	194
73.	Oyster-catcher	H. W. Richmond	196
74.	Oyster-catcher's eggs	,,	197
75.	Avocets	,,	198
76.	Woodcock	,,	200
77.	Common Snipe	J. Holmes	202
78.	Ruff	H. W. Richmond	209
79.	Redshank's nest and eggs	T. L. Smith	213
80.	Curlew on nest	R. Ll. Bruce	217
81.	Common Tern	H. W. Richmond	224
82.	Herring-Gull	,,	228
83.	Guillemots and Kittiwakes on the Farne Islands	,,	231

			PAGE
84.	Arctic Skua on its eggs	H. W. Richmond	234
85.	Great Auk, from the Cambridge University specimen		237
86.	Guillemots on the "Pinnacles," Farne Islands	H. W. Richmond	239
87.	Black Guillemots.	,,	240
88.	Puffins on Farne Islands	,,	242
89.	Red-throated Diver on its eggs	,,	245
90.	Great Crested Grebe on nest	K. J. A. Davis	247
91.	Little Grebe's nest	T. L. Smith	249
92.	Little Grebe's nest with eggs covered	,,	249
93.	Storm-Petrel	H. W. Richmond	251
94.	Fulmars on eggs	,,	255

These illustrations, of which all but the first three are original, have been reproduced from photographs, taken for the most part from Nature, but in some cases at the Zoological Gardens or elsewhere. It is my pleasant duty to offer my thanks to the many kind friends who have assisted in this work, especially Mr H. W. Richmond of King's College, Cambridge, the authorities of the Zoological Society, and those of the Natural History Department of the British Museum.

INTRODUCTION

ON THE CLASS *AVES*, OR BIRDS IN GENERAL

Birds are distinguished from all other living creatures by their covering of feathers. They are moreover bipeds, and have beaks, wings, and tails; but these features are not peculiar to them, while the power of song and the method of reproduction by eggs are also held in common with other animals. Again, ability to fly, in the true sense of the term, is a possession of most species at the present day, but Palæontology teaches us that of old there were flying Lizards, and even now we have flying mammals in the shape of Bats. None of these points therefore, with the exception of the first, are unfailing characteristics of the Class *Aves*.

There is no doubt that the ancestors of our birds bore a remarkable resemblance to reptiles, and that, if they did not actually spring from them, as is now the orthodox belief, both must have certainly arisen from a common origin, that is, from some creatures combining in themselves those points which the two classes have in common. This is the more evident when we consider the earliest known fossil bird, now termed *Archæopteryx lithographica*, which was discovered at Solenhofen in the kingdom of Bavaria. It was about the size of a Rook, and was in all probability a tree-loving species.

Archæopteryx lithographica

It had teeth in both jaws of the short blunt beak, a long lizard-like tail with twelve big feathers on each side, wings with both primary and secondary quills, and perhaps a weak keel to the breast-bone. The first two of these points, with other features of the skeleton, distinguish it so sharply from all other birds that it has been placed in a Subclass by itself, termed *Archæornithes* (ancient birds).

Certain of the latter-day birds have no keel to the breast-bone and therefore no attachment for the flight-muscles—for instance Ostriches, Emus, and Cassowaries; these have been separated from the others and termed *Ratitæ*, as opposed to forms with more or less of a keel, which are known as *Carinatæ*, the latter including all our British species, as will be seen below.

We cannot here deal with osteology or anatomy, but the subjoined figure will explain the technical terms used for the feathers on different parts of a bird by scientific writers. They do not grow on all parts of the body alike, but on certain tracts called *pterylæ*, while the unfeathered parts are named *apteria*. It may be of interest to our readers to learn that the sprouting feather consists of a "barrel" or quill which bears a tuft of rays called barbs, and that these again by splitting ordinarily produce "barbules." The earliest and softest feathers are those which are collectively called "down," while below each down-feather is formed a contour or webbed feather, so that eventually the latter protrudes with the former at its tip. Subsequently this falls off, as may easily be observed in the young of any down-clad species. Many nestlings, however, have little or no down at all; on the other hand many down-feathers remain continually in their original state. In contour-feathers the barbules give rise to " barbicels " which regularly end in little hooks

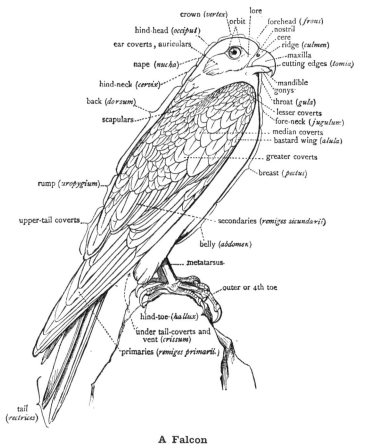

A Falcon
To shew the nomenclature of the external parts

that catch in the folded margins of the next row and serve to produce a firm surface or " web " on each side

Introduction

of the shaft. If barbicels, or even barbules, are absent, the feather is called decomposed; if the barbs also are wanting, we have bare quills, wires or bristles.

Birds do not, however, perpetually keep the same coat of feathers, but have periodical "moults," or annual changes. The young do not always lose their main quills in their first year; on the other hand certain groups of birds not only pass through a regular autumnal moult, but have a second change of the smaller feathers in the following spring. These processes account for the difference between the summer and winter plumages, while some species are known to have three moults, and therefore a distinct summer, autumn, and winter coloration. Decorative plumes are generally assumed in the spring, and are chiefly to be found among the males, which are in the great majority of cases brighter—and larger—than the females. The young are generally similar to the female in colour before they moult.

There are curious exceptions to the above rules, for instance in the Phalaropes and Hemipodes the female is the larger and brighter bird—and there the male takes her place in courtship and incubation; in Penguins the feathers flake off, instead of moulting properly; Gannets take some six years to attain their full adult plumage; most of the Duck tribe lose all their wing-quills at once and then become flightless, while the males temporarily assume the plumage of the females and are said to suffer "eclipse."

The colour of a bird's plumage varies with the seasonal moults, for the most part by new feathers taking the place of the old, but also by the partial or entire wearing away of the edges, whereas change of the actual colouring matter, if indeed it occurs at all, must be most exceptional. Gloss and iridescence are due to the structure of the feathers, which may be polished or shew little ridges under the microscope.

Newly hatched birds often run from the shell, and are therefore called "nidifugous" or "nest-deserting," but the majority are "nidicolous" or "nest-inhabiting," and fly only when fully fledged; some again of those which run from the shell can only move to short distances for a considerable time. These habits are naturally related to the position of the nest.

With the exceptions already mentioned, and excluding the time of moult, birds have extraordinary powers of flight, though these powers are in constant use only in the case of certain forms, and in others are put forth periodically; speed, endurance, and like factors here come to be considered, while the style of movement, including the amount of wing-action or oarage, and tail-action or steerage power, varies to an enormous extent. The greatest example of untiring flight is that of an Albatros, which will follow a ship for days together; but the same habit, to a less extent, must to most people be familiar in the Gull tribe, which also follow vessels for long periods, and remain on the wing for hours, when looking out for food. These birds are not always flapping their wings, but glide or skim along with intervals of muscular action; while it is evident that they are greatly indebted to the supporting power of the air and its force against the flight-feathers. Vultures, Eagles and Buzzards, Falcons and Hawks have a different kind of flight; they either cleave the air at a great pace or move along slowly but powerfully; and even if they soar and circle round in the sky, or hover, as in the case of the Kestrel, their wings are for the most part in a state of constant motion. Many of this family are noted for the swiftness with which they dash upon their prey, so different, for instance, from the soft noiseless progress of an Owl. Storks and Cranes on migration fly for huge distances at great elevations; Swans, Geese, and Ducks, heavy creatures though they are, move at a pace that is readily admitted by the gunner who misses them; Pheasants, Partridges, and Grouse can travel at great rates, and the first-named rise with amazing suddenness from the ground; among the Plovers the Lapwing is noted for its noisy "winnowing" flight, partly due to the wide expanse of its wings; Woodpeckers follow an undulating course, Kingfishers dart from place to place, Larks soar, and almost every group of birds has different methods, of which these are but a few striking examples.

Again, the smallest species traverse incredible distances on migration. This is not perhaps so marvellous in the case of the Swallow, which careers through the air in untiring fashion at any time, and only alights at intervals, nor in the case of the more powerful Swift, which seems to be able to remain aloft indefinitely, and is rarely seen to perch except at the nest or when roosting; but it is astonishing beyond measure in birds of apparently limited flight, such as Thrushes, Wagtails, Pipits, and dozens of others. Of these the Golden-crested Wren is perhaps the most wonderful, for though the weight to be supported is here inconsiderable, the delicate structure appears to be little fitted to cope with the stormy weather that is often prevalent at the seasons of passage.

The mechanics of flight are, of course, far beyond our scope, nor will we attempt to account for all the different modes of progression, but we may call attention to the fact that the power of locomotion does not depend on length or strength of wing alone, while the shape of a bird's body, which is often provided with air-sacs, and the more or less hollow bones, are well calculated to make progress in the air as easy as possible.

Consideration of flight naturally leads us to the subject of Migration, which, indeed, we have already been obliged to mention. From ancient times it has been one of the marvels of bird-life, being referred to in the book of Job, by Homer, and the later Greek and Latin poets, as well as by countless subsequent writers. Yet we seem to understand the phenomenon only slightly better than of old, though great efforts have been made of late years to gain more definite knowledge of the magnitude and direction of the movements.

In the first place it is necessary to mention three classes of birds which are often confounded under the name of migrants. The true or summer migrants, as regards Britain, are those which, after breeding in our country, leave it for the winter and return again in spring, the times of their arrival and departure being more or less variable quantities. The partial migrants on the other hand are those which may be said to be more or less resident in Britain as species, though many individuals leave us on migration, and many that have not bred with us visit us at the colder seasons. The birds of passage are those which are only seen for a shorter or longer period in autumn, pass on to more genial climates, and are frequently again in evidence for similar periods in spring. The term resident is applied not only to species which never or very exceptionally leave us, but also to those which are for the greater part non-migratory or only share in limited movements within the kingdom. The word "resident" in fact is often used in a comparative sense, and many birds of this description are really partial migrants; this must necessarily be the case as long as we cannot safely assert that the individuals met with in winter are the same as those breeding with us in summer.

Many attempts have been made to ascertain the distances travelled on migration and the direction followed by the flocks, as well as their numbers, the altitude of flight, the pace in the several cases, the most favourable weather, and so forth. Much has been discovered with regard to the four last points by continued observations at Lighthouses and Observatories, coupled with those made on the rate of flight of individual birds; but much less success has attended the constant efforts to determine the two first points. Mr Eagle Clarke in particular has spent an immense amount of time at the seasons of migration at Lighthouses, or on Lightships; the keepers of the Lighthouses have aided by transmitting specimens that have been killed at the Lights from many quarters, while Mr Clarke has prepared an abstract of such reports; marked rings have been fastened to birds' legs at the nesting places by ornithologists in different countries with a view to ascertaining where they occur at later periods; and finally watchers have noted the arrival and departure of the different species and filled lengthy lists with their observations. Yet all this good work has but resulted in confirming what was already pretty well understood—that is, the great distances traversed by certain birds, and the general direction of their movements. On the other hand a mass of the most valuable information has been accumulated with regard to the methods of migration. Species which ordinarily travel in flocks can be separated from those which are apt to do so singly or in pairs; the numbers in the flocks have been proved often to be incalculable; the altitude has been reckoned in certain cases and found to be so great that it is clear that only the lower flocks are really brought within our ken; bad weather has not proved to be in all cases an obstacle to migration, though the direction of the wind has always to be considered.

Apart from the distances traversed, the direction of migration, that is, the broad lines in which different species travel, is a question of great moment. Birds which breed to the south of the equator certainly tend to migrate northwards; but so little is known of the habits of these southern forms that we must follow the course, usual at present, of confining our remarks to those that breed in the northern hemisphere, while noting that the movements are of much less extent in the southern half of the globe.

On migration the young usually start before the parents, though in exceptional cases, such as that of the Cuckoo, which is reared more often than not by one of our resident species, they linger till a later date. Once started, the direction is distinctly influenced by the conformation of the land; coasts, river-valleys, and so forth making for ease of travel, high mountain ranges for difficulty; but even the last-named are not uncommonly surmounted, and the old idea that straight lines were more or less followed has, in the minds of most people, given way to the certainty that the flocks gradually spread over large areas, and that considerable deviations occur, for which at present it seems impossible to account. Many birds, such as our Swallows, are seen to collect together some time before they leave us, others, such as Woodcocks, arrive simultaneously in large flocks, while close observation soon shews that a very great number of other species act similarly, and that "rushes" continually occur, which are most strikingly witnessed at Lighthouses. Migration, however, takes place largely at night.

The causes of migration have been a fruitful source of discussion, but there is a general agreement that changes of temperature and the available food-supply are the most effective. Exceptionally hardy birds, such as Penguins in the southern oceans and the Spitsbergen Ptarmigan in the north, especially if they live on the sea or in districts thinly populated by their kind, need hardly migrate at all; but, as regards the more delicate forms, the colder weather that begins in autumn might of itself be sufficient to drive the birds from their breeding quarters. This colder weather also diminishes the supply of insect food, while the season of berries and other fruits soon comes to an end, and even the smaller mammals on which some birds feed almost cease to be seen. It does not follow that northern species stand in need of very high temperatures during the winter; if so, they would probably stay in great numbers near the equator, instead of so constantly passing further to the south, and, as already stated, the more adaptable species, specialized to that effect, may remain throughout the year in Arctic or

Antarctic climates. All birds that fly are capable of migration to a greater or less extent, but all do not take full advantage of their capabilities.

The return migration in spring to the northerly breeding haunts may be due to some hereditary instinct —whatever that expression may mean—or to the fact that the birds seek what they know to be the places most favourable for rearing their young in the particular manner to which they are accustomed, or even to the pressure of the species of the southern hemisphere which may be returning from their winter quarters, though this argument does not seem very convincing.

It has also been suggested that migration may be due to the same cause as an extension of breeding range; that is, the numbers may become too great to be contained in the original summer haunts, and under pressure a certain proportion may move further to the south, when they cannot well go further northward. But, since it can hardly be contended that there is absolutely no room left in most areas for the birds' nests, and since so many of them breed in colonies and do not at present dispute the possession of every inch of soil, this argument practically resolves itself into that concerned with scarcity of food.

Finally it may be observed that the length of the journeys taken by various species of birds differs immensely; many move to comparatively short distances, while such forms as the Turnstone and the Sanderling, which breed in the far north, go so far south in winter that they may practically be considered cosmopolitan.

Migration naturally leads to thoughts of geographical distribution, and care must be taken not to confound the two questions. By distribution we mean, unless it is otherwise stated, the limits within which a species ranges in the breeding season, just as the bird's range is, if unqualified, taken to mean its summer range. Even in Britain distribution is often limited, as in the case of the Snow Bunting of the Scottish highlands or the Nightingale of England; but the range of each species will be found given in detail below, and need not delay us here. Many birds throughout the world are extremely local; many are only found in deserts, isolated islands, and so forth, while the Red Grouse may be given as the example nearest to our doors, as it is absolutely confined, as a native species, to the United Kingdom. But Britain is a comparatively small area, and it is necessary to survey the whole globe in distributional questions. This is far beyond our scope, and we need only state that, taking into consideration birds alone, Dr P. L. Sclater suggested a division of the world into six Regions, the Palæarctic, Ethiopian, Indian, Australian, Nearctic, and Neotropical (see map), in each of which he considered the forms—taken as a whole— to be more closely connected with each other than with those of another Region. The Palæarctic and Nearctic (of the Old and the New Worlds) together form the Holarctic; New Zealand may be considered separately.

If our readers ever proceed from the study of British Birds to that of foreign species, they will be greatly struck by the prevalence of very peculiar forms in certain countries, by the wonders of the Arctic and Antarctic areas, by the marvels of desert and island life, by curious extinct birds, by extraordinary habits of courtship, sexual display, parasitism, and many other subjects, but even within our islands they may spend a pleasant and useful lifetime in observing the species they meet with, in protecting them and carefully noting their habits, which will often be found to be much more peculiar than is usually supposed to be the case. Their cleverness is not uncommonly remarkable, and their structure admirably adapted to the needs of their

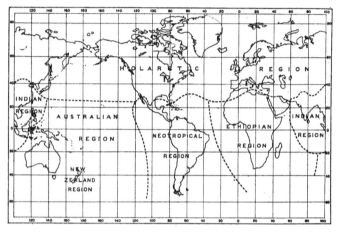

The Zoo-Geographical Regions

existence, while the changes of plumage are of themselves a lifelong study and are none too well understood even at the present day, especially with regard to the young bird. No doubt all the species on our list cannot be found in any one locality, but there is no district where much good work may not be done.

CLASSIFICATION

Birds may roughly be classed as land birds and water birds, but there are many that can hardly be included properly in either category; it is still less possible to divide them by their dwelling-places, as those of the woods, fields, shore, and so forth; we require some formal classification into "Families" under the larger groups called "Orders," though it is clear that no linear arrangement can be entirely satisfactory, or will shew all the relations between the members of the "class" *Aves*. We will therefore give here the scheme used in the following pages, which is almost the same as that in Howard Saunders' *Manual of British Birds*. The nomenclature is almost exactly that of the new List of the British Ornithologists' Union.

AVES CARINATÆ

ORDER I. PASSERES

Family TURDIDÆ
  Subfamily *Turdinæ*
  Subfamily *Sylviinæ*
  Subfamily *Accentorinæ*
Family CINCLIDÆ
Family PANURIDÆ
Family PARIDÆ
Family REGULIDÆ
Family SITTIDÆ
Family TROGLODYTIDÆ
Family CERTHIIDÆ
Family MOTACILLIDÆ

Family ORIOLIDÆ
Family LANIIDÆ
Family AMPELIDÆ
Family MUSCICAPIDÆ
Family HIRUNDINIDÆ
Family FRINGILLIDÆ
  Subfamily *Fringillinæ*
  Subfamily *Emberizinæ*
Family STURNIDÆ
Family CORVIDÆ
Family ALAUDIDÆ

ORDER II. PICARIÆ

Family CYPSELIDÆ
Family CAPRIMULGIDÆ
Family PICIDÆ
  Subfamily *Iynginæ*
  Subfamily *Picinæ*

Family ALCEDINIDÆ
Family CORACIIDÆ
Family UPUPIDÆ
Family CUCULIDÆ

ORDER III. STRIGES

Family STRIGIDÆ

ORDER IV. ACCIPITRES

Family VULTURIDÆ
Family FALCONIDÆ

ORDER V. STEGANOPODES

Family PELECANIDÆ

ORDER VI. HERODIONES

Family ARDEIDÆ
Family CICONIIDÆ

Family IBIDIDÆ
Family PLATALEIDÆ

ORDER VII. ODONTOGLOSSÆ

Family PHŒNICOPTERIDÆ

ORDER VIII. ANSERES

Family ANATIDÆ
  Subfamily *Anserinæ*
  Subfamily *Cygninæ*
  Subfamily *Anatinæ*
  Subfamily *Fuligulinæ*
  Subfamily *Merginæ*

ORDER IX. COLUMBÆ

Family COLUMBIDÆ

ORDER X. PTEROCLETES

Family PTEROCLIDÆ

ORDER XI. GALLINÆ

Family TETRAONIDÆ

Family PHASIANIDÆ

ORDER XII. GRALLÆ

Suborder **Fulicariæ**
Family RALLIDÆ
Suborder **Grues**
Family GRUIDÆ

Suborder **Otides**
Family OTIDIDÆ

ORDER XIII. LIMICOLÆ

Family ŒDICNEMIDÆ
Family GLAREOLIDÆ

Family CHARADRIIDÆ

ORDER XIV. GAVIÆ

Family LARIDÆ
  Subfamily *Sterninæ*
  Subfamily *Larinæ*

Family STERCORARIIDÆ

ORDER XV. ALCÆ

Family ALCIDÆ
  Subfamily *Alcinæ*
  Subfamily *Fraterculinæ*

ORDER XVI. PYGOPODES

Family COLYMBIDÆ

Family PODICIPEDIDÆ

ORDER XVII. TUBINARES

Family PROCELLARIIDÆ
Family PUFFINIDÆ

Family DIOMEDEIDÆ

ORDER I. PASSERES

This Order includes all the true singing birds, in which the vocal organs are most highly developed; moreover in every respect they certainly represent the highest stage of development in the Class. The members are of all sizes and colours, and for the most part haunt trees or bushes, so that they have been somewhat arbitrarily called perching birds, though perching is not a habit peculiar to them. The toes are all on the same level and never webbed; the young are born naked and helpless; but these characteristics will not alone determine the Order, which contains about half of the birds that are at present known.

It may be well to mention three important points before going further, to prevent future confusion. First, the descriptions of the Orders must be taken to refer to British forms alone; second, the range of a bird is used in the sense of its breeding range; third, what is commonly termed a bird's leg is mainly its foot, the real leg being often hidden by the plumage. The word foot is hereafter used in its correct sense.

Family TURDIDÆ, or Thrushes, Warblers, and their Allies

SUBFAMILY **Turdinæ**, OR THRUSHES, CHATS, REDSTARTS, BLUETHROATS, ROBINS, AND NIGHTINGALES

This Subfamily is very closely allied to that of the Warblers (*Sylviinæ*), but differs in the more or less spotted condition of the young, above and below, until their autumn moult; of this the Robin is a good instance.

Three typical Thrushes are residents or partial migrants in Britain. Of these the Mistle, *i.e.* Mistletoe-thrush (*Turdus viscivorus*), is the largest, and is especially noticeable in early spring, when its loud churring notes, coupled with a harsher song than that of the Common Thrush, may be heard in the roughest weather. Hence it is known as the Stormcock, while it is often called a Feltyfare in mistake for the Fieldfare. In plumage it is greyer than the common thrush, with white instead of orange-buff under-wing; in flight it is swifter; in food it shews a greater liking for fruit and berries, including those of the mistletoe. The young have much white on the wing-coverts. The Mistle-thrush is a bird of the hedgerow and copse, not of thick woods, and has spread to the northern islands of Scotland and to Ireland during the last century or so, with the increase of plantations. The nest, a solid structure of grass and moss, lined with finer materials, is seldom placed near the ground, but occupies some conspicuous fork of a branch; the four or five greenish or reddish white eggs are finely marked with rusty red and lilac. This species is an early breeder, but is certainly to a large extent migratory; it is a shy bird, though bold at the nest. It occurs throughout most of the Palæarctic region, but not within the Arctic circle.

The Common or Song-thrush (*T. musicus clarkii*) is too familiar to need description either as regards its plumage or its varied song. Its foreign range is not very dissimilar to that of the last species, but the continental form is distinguishable from the British in coloration, and we are thus able to gain a rough idea of the number of individuals that arrive from abroad in autumn or leave us for the winter, though some are undoubtedly resident. With us it ranges even to Shetland, and occurs fairly high on the hills, the well-known mud-lined nest being occasionally placed on the ground, but generally in trees and shrubs. It lays from four to six blue eggs with black or brownish spots. Except during the moult the song may be heard at any season, while the bird has its first brood exceptionally early in the year. The food consists of berries and insects, worms, slugs, and snails, especially the last, which are usually smashed on some favourite stone.

The Blackbird (*T. merula*) is so called from the cock, which is black with orange bill, while everyone knows that the hen and young are dull brown with the bill dusky. When feeding on the ground this species has not quite the " hop and run " action of the thrush, but its food is the same and its flight similar. It is, however, a bird of lower levels, constantly flushed from hedgerows or bushes, and less often sitting on tree-tops to sing. The song is comparatively flute-like and mellow. The nest of dry grass and mud, with grass lining, contains about five green eggs with small rufous markings, and is placed at no great height from the ground or even on it. Common throughout Britain and a partial migrant, the Blackbird is not found outside Europe, except in Asia Minor, Palestine, and north-west Africa, with the Atlantic islands.

In the hilly moorlands from Cornwall and Wales to northern England and Scotland it is represented by the Ring-ousel or Hill-blackbird (*T. torquatus*), characterized by the white chest, less conspicuous in the female and still less in the young, and also by the yellowish bill with black tip. This migratory species arrives about April and leaves us by October, with the exception of belated individuals, while it only breeds in northern Europe, the central and southern European form being distinguishable. Its ringing note or whistle is a characteristic sound of our higher hill-slopes, but the performer is less often seen than heard, for it is a shy bird except at the nest. This is like that of the Blackbird, but is placed among rocks, in heather, or

Ring-ousel

on banks of streamlets; the eggs are brighter both in the ground-colour and the larger spots. Berries form a considerable part of the food, especially on passage, which the birds compass easily.

We now come to two migratory species, common with us in winter, the Redwing and the Fieldfare, which somewhat resemble a small Song-thrush and Mistle-thrush respectively. The Redwing (*T. iliacus*) may, however, be distinguished by the red, in place of orange-buff, below the wing and the broad whitish stripe above the eye; the Fieldfare (*T. pilaris*) by the distinctly grey head and rump and the yellower breast. They both breed in northern Europe and in Asia eastward to the Lena river, but the former alone nests in Iceland and the Færoes, while the latter ranges further south to central Germany and Austria-Hungary, and often forms colonies. The nests and eggs are much like those of the Blackbird. These Thrushes come in large flocks in October or earlier, and leave about April; the Fieldfares keeping more together and being easily recognisable by their harsh chattering cries as they fly about the fields and hedgerows, whereas the Redwings separate and have a more melodious call. They are shy birds of swift flight, often shot for the table in winter, and not uncommonly killed by severe cold, the Redwing being the first to succumb to the effect of the weather.

Very different in appearance from the true Thrushes are the Chats, Redstarts and Bluethroats, the Redbreast and the Nightingale; nevertheless they are the connecting links with the Warblers, which are usually on anatomical and other grounds included in the Family *Turdidæ* as a Subfamily *Sylviinæ*. The habits vary, but the food seems always, if we leave the Robin out of consideration, to consist of insects and their larvæ, spiders, worms, and small mollusks. The Chats are conspicuous and lively birds, with jerky flight, pleasant little songs, and sharp clinking call-notes that explain their name, the largest being the Wheatear, *i.e.* white-rump (*Œnanthe œnanthe*), which is locally abundant on moorlands, downs, and sandy warrens. It is one of our earliest spring migrants, arriving in mid-March, and making pretty straight for its breeding quarters, while it leaves the country by October. But it is not till April or May that it builds its nest of grass lined with fur, hair, or feathers, which is placed in a rabbit-burrow or similar excavation in most cases, though it may be

Wheatear

in holes in walls or peat-stacks on the moors, while a little building material at the entrance generally betrays the site. The five or six eggs are very pale blue, rarely with a few purplish specks. The Wheatear, from its similar note, is often confounded with the Stonechat, but may readily be distinguished by its grey back, black cheeks, wings, and tail, white rump and breast.

The female is browner and buff below. The range extends over the whole Palæarctic region from Jan Mayen and Mongolia southwards and even to the Azores, but a larger race inhabits north-east America and Greenland and visits Britain on migration.

The Whinchat (*Saxicola rubetra*) is found in rough grassy places of various descriptions, with a preference for moors and newly planted copses; it is mottled with brown and buff above and is fawn-coloured below, having over the eye a distinct white streak—which is buff in the female—and some white on the wing and tail. The nest, usually placed near the base of a small shrub or large herbaceous plant, is a mossy structure with a lining of fine grass, and contains about six green-blue eggs, generally with rufous spotting. The hen-bird sits very closely, while when disturbed both parents flit before the intruder, perching on the shrubs, and repeatedly uttering their alarm note of "u-tick." Breeding takes place about mid-May, but the bird arrives a month earlier and stays till October. Abroad it ranges through Europe to west Siberia, though it keeps to the hill country in the south.

The Stonechat (*S. rubicola*) should really be called the Whinchat or Furzechat, as it is most common among furze, where it may be seen throughout the year, though as a species it is partly migratory. The cock, a brown bird with black head, a ruddy breast, a white patch on the wing and a partial white collar, is very conspicuous as it flits in a fussy way from one perch to another, uttering its clicking notes as a warning to the hen, which is almost brown. The nest is placed among heather, rough grass, or very low gorse, and is made of moss, grass and so forth, with a finer lining;

the five or six eggs are greenish with rufous spots, usually placed in a ring. They may be found from early April to August, so doubtless two broods are reared in a season. The nest is well concealed and the hen sits closely. Our bird is confined to Europe and north Africa, but has several near relatives.

Redstart

The Redstart or Fire-tail (*Phœnicurus phœnicurus*) is well known throughout England and Scotland from April to September, but is rare in Ireland and hardly reaches our northern isles. It is by no means shy and often very bold at the nest, while a bird of active ways with grey back, white forehead, black face and throat, chestnut breast, and orange-red rump and tail, which it is always flirting, is not easily overlooked, even in the wooded country which it chiefly haunts. Its sweet low warble is perhaps most commonly heard in the valleys of our northern hills; the nest of moss, grass, and roots, lined with hair and feathers, is generally built in a hole in a rotten tree or wall, and contains some six pale blue eggs. The bird has a wide foreign range from northern Norway to Lake Baikal, and southward throughout Europe to the Atlas mountains in north Africa, but complications arise from the existence of several closely allied forms. The hen has a plain head and brownish back.

The Black Redstart (*P. titys*), as its name implies, is black with a red tail, the upper surface being somewhat greyer with a white wing-patch and the two central tail feathers brown, as they are in the Common Redstart. To travellers in Germany it is a well-known bird, for it builds its nest round houses and sheds, while it is specially interesting to us as having been suspected of breeding in England and being a frequent autumn visitor. The female resembles that of the last species but is greyer; the male has a richer song; the eggs are white. The range extends from the Baltic and the Urals to Rumania, Palestine, and north Africa.

The Red-spotted Bluethroat (*Cyanosylvia suecica*) is an irregular autumn and rare spring migrant, which deserves special notice as linking the Redstarts to the Robin and the Nightingale, and so to the Warblers. It has the general habits and nest of the Robin, while the song is little inferior to that of the Nightingale, and the eggs are similar. The male is brown, with white eye-stripe, rufous rump, and blue throat; the throat has a red central spot and is bounded by black, followed by a rufous patch above the white belly: the female has little blue on the throat and a brown chest-band. This species only breeds in the north of Europe and Asia, but there is a form with a white instead of a red spot which carries on the range to France and west Russia.

Our familiar Robin or Redbreast (*Erithacus rubecula melophilus*) needs no description; its plumage, habits,

Robin

and song are equally well known. It may be mentioned, however, that it is not known to breed in the Færoes or Shetlands, though it ranges from north Europe and west Asia to Africa and the Atlantic islands, while the continental race is clearly distinguishable and visits us in winter and the African form has also been separated. Holes in banks, walls and trees are utilized for the nest of leaves and moss lined with hair, while the bird has a great fancy for an old can or box, as occasionally happens with the Wheatear and Stonechat. The first brood may be hatched early in April; the eggs are white with rufous spots.

The Nightingale (*Luscinia megarhyncha*) is no doubt our most wonderful songster, though it is approached by the Thrush and nearly equalled by the Blackcap and

Nightingale

the Garden Warbler. Its song, however, gives way to a harsh churr when the young are hatched, as in the case of so many Warblers. Arriving in April it soon becomes common in eastern England, though less abundant westward, and barely known in Devon, Hereford and Cheshire. Exceptionally it has been found breeding in Glamorgan and north Yorkshire, as well as on one occasion in Northumberland. Though the Nightingale remains with us till September, it is little seen after the breeding season, and even then chiefly as a reddish brown bird which pops out in front of the observer and flies along to a neighbouring hedge or bush. The curious nest is made outside of dry leaves, usually of the oak, and is placed in vegetation on the ground, or close to it; the five or six eggs are olive coloured or greenish with olive-brown markings. The range extends from central Europe to Asia Minor and north Africa.

SUBFAMILY **Sylviinæ**, OR WARBLERS

In treating of the Warblers, the first point to notice is that all students of bird-life must learn their notes in the field. We cannot pretend to reproduce them here, and the syllables given to represent them in books are rarely understood alike by any two persons. This is, moreover, true of most bird-voices; we may truthfully talk of a croak, a click, a hoot, a warble, and so forth, or speak of such as harsh or sweet, but attempts to imitate them on paper are sure to mislead. Warblers live on slugs, worms, spiders, insects—aquatic or otherwise—and their larvæ, according to the species, and some on fruits also; insects are often captured on the wing, while the smaller members of the group flit characteristically among the leaves of trees, hunting for their prey.

With this prelude we may proceed to the Common Whitethroat (*Sylvia communis*), the familiar species of our more open woods and roadsides, easily recognised by its early arrival, its habit of springing up a few feet in the air to utter its monotonous notes before settling again on the hedgerow, its white throat and its ruddy brown colour above. The more shy Lesser Whitethroat (*S. curruca*) has a white instead of a buff breast and a very much finer song of a somewhat similar character; it is always a more local bird, not nesting in Ireland, very seldom north of the Border and sparsely in Wales, while the commoner species is universal south of Sutherland and Caithness. Similarly, outside of Britain, the range of both covers Europe south of lat. 65° N. and southwest Asia, though in some parts the smaller bird is the more abundant. Moreover it extends to Siberia, while its congener breeds in Algeria and Tunisia; but all depends on the number of forms we recognise, as several are closely allied. The flimsy nest of grass-stems and cleavers is placed in low bushes, brambles and hedges; but the Common Whitethroat is fond of nettles and coarse herbage to conceal its nursery, while that of the Lesser is more often in young hawthorns, blackthorns, and similar situations, and seems absurdly small in comparison. Its five or six eggs, moreover, have a clearer white ground than those of its congener, with olive and brown rather than green markings, and resemble those of the larger Garden Warbler.

The Blackcap Warbler (*S. atricapilla*) and the Garden Warbler (*S. simplex*) are in song worthy rivals of the Nightingale, which some may even consider inferior, though it certainly has more variety of phrases. But, while the songs of the two species may be easily mistaken, the male Blackcap with its black crown, grey nape and under parts can never be taken for the olive-brown Garden Warbler with its whitish lower surface. The female, which has a red-brown instead of a black head, is particularly hard to identify in a dark thicket; yet this is often necessary, as the nests are similar and the whitish eggs with yellowish brown spots may correspond exactly. It is generally possible to make a correct guess at the nest, which in the Garden Warbler is less flimsy and better lined with hair, but the very green variety of its eggs and the rare red variety of those of the Blackcap can alone be guaranteed without a sight of the parent. All the Warblers so far mentioned arrive about April and leave us in September, but the Common Whitethroat is the earliest and the Garden Warbler the latest. The Blackcap and Garden Warblers both range in Britain up to mid-Scotland, and the former a little further north; they vary in abundance locally, but agree in being scarce in Ireland. They breed over Europe, except the more Arctic portions, and in north-west Africa, but the eastern limits in Asia seem to lie in western Siberia, and only the Blackcap nests in the Atlantic islands.

The Dartford Warbler (*Melizophilus undatus*) is a jolly little dark grey bird with chestnut breast, which is usually seen flitting restlessly about the gorse bushes or tall heather; it is now scarcer with us than formerly, and is confined to East Anglia, Shropshire, and the south of England. Abroad it occurs in north-west France and in barely separable forms to Italy, Morocco, and Algeria. Though skulking at other times this local resident is often bold enough in the breeding season, when the cock utters his scolding notes from a spray just ahead of the intruder, and shifts his quarters but slightly when disturbed. The nest and eggs much resemble those of the Whitethroat, but the latter are somewhat longer with more olive or even reddish markings. The site of the nest is in a gorse bush or heather clump, while in winter the birds move from place to place for shelter, as they are the reverse of hardy. The foreign range is complicated by the presence in the Mediterranean of the more uniformly grey Marmora's Warbler.

Willow Wren

The Willow Warbler or Willow "Wren" (*Phylloscopus trochilus*) is the commonest of three little yellowish Warblers which appear very early in spring and leave under ordinary circumstances in September. It is no less abundant in the north than in the south and in Ireland, and even breeds in Shetland, while abroad the typical form ranges from Lapland and north Russia to the Mediterranean, except for the Balkan Peninsula.

The Chiffchaff (*P. collybita*), a more local bird, which barely extends to northern Scotland, is only found in its typical form from France and Germany southward in Europe and is replaced east of the Volga by the Siberian Chiffchaff, known also to breed in north Russia. The Scandinavian Chiffchaff occupies north and east Europe, and a smaller form is found in the Canaries. The Wood Wren (*P. sibilatrix*) is the rarest of our three species, and prefers oak woods and hill-valleys; it is a somewhat less northern bird than the Willow Wren, but reaches the Mediterranean. All three are yellowish green above and lighter below, with a yellow streak over the eye, but the distinctly yellow breast of the Wood Wren clearly distinguishes it from the other two, in which the breast is yellowish white. The Chiffchaff is duller and smaller than the Willow Wren with darker feet. But, apart from plumage, they are easily recognisable by their notes and breeding habits. The song of the Willow Wren consists of a few sweet reiterated notes, occasionally swelling into a song like that of the Garden Warbler; the Wood Wren, after a similar start, ends with a long drawn trill; the Chiffchaff says "chiff-chaff" most distinctly. The song is usually accompanied by a quivering of the wings: the hens merely utter a plaintive sound. The nests are oval balls of grass with a side entrance; but those of the Willow Wren and Wood Wren are very seldom placed above the ground, while that of the Chiffchaff is nearly always in a low bramble, small shrub, fern or grass-tuft; the Wood Wren lacks the usual feather lining and has five or six eggs thickly spotted with purplish black, the spots in the Chiffchaff being also purplish though sparse, and in the Willow Wren red. The last-named does not coat its nest with dry leaves, as the others constantly do. These three birds are regularly seen foraging over the green leaves on the higher branches of trees for flies, aphides, and the like, a habit which has given them the name of *Phylloscopus* or " leaf-investigator "; the Chiffchaff is the earliest to arrive, the Wood Wren the latest.

We now come to another small group of three migratory Warblers, which visit us between the latter part of April and September; their general coloration is reddish brown with buffish white under parts, and they have a conspicuously rounded tail. Careful attention is necessary on the part of a beginner to distinguish them by their hurried babbling notes, apart from difficulties with regard to the plumage. All are aquatic, but the Reed Warbler (*Acrocephalus streperus*) is seldom found except in beds of reeds (*Phragmitis vulgaris*) and does not range north of Yorkshire, or to Ireland: abroad it occurs from south Sweden and south-western Siberia down to the Mediterranean, north-west Africa, and Baluchistan. The Marsh Warbler (*A. palustris*) has a slightly more southern foreign range, and is at present only recorded as breeding with us in Somerset, Gloucester, Worcester, Oxford, Wilts, Hants, Sussex, Kent, Surrey, Bucks, Cambs, and Norfolk; it usually haunts osier holts and damp copses, and eschews reed-beds. The Sedge Warbler (*A. schœnobœnus*) breeds throughout Britain in suitable places, but always near water, though a small ditch may suffice, whereas the last-named species has been known to nest in cornfields.

Abroad the Sedge Warbler extends throughout Europe from 68°—70° N. lat. to the Mediterranean, and westward in Asia to north Siberia and the Altai mountains.

Reed Warbler's nest

Experts themselves often fail to distinguish skins of the Reed and Marsh Warblers, but the latter is more olive above and has brownish flesh-coloured in place of purplish brown legs. Both these birds are nearly white below and have an obscure buff streak over the eye: the Sedge Warbler on the contrary has a buff breast, a distinct yellowish white eye-streak, and a streaked instead of a plain crown. The respective haunts and habits are, however, characteristic, while the birds, though in constant motion except when singing, can scarcely be called shy and are easily observed. When not built in herbaceous plants or bushes, the Reed Warbler's nest is slung between three or four reeds standing in water, and is an elongated structure with a deep cup, lined with wool or hair, to contain the four or five greenish white eggs with olive and grey blotching; it almost dwindles to a point below, and is a curiously hard-looking fabric, which, it may be remarked, is one of the most favourite nurseries of the Cuckoo. The Marsh Warbler seems independent of water, except that, in the same way as its congeners, it relies largely for food on aquatic insects and their larvæ, while the flatter nest is never in reeds but is usually placed in rough herbage or small willows, meadow-sweet and willow-herb being specially favoured sites. The eggs are distinguishable by their white ground-colour and clearer markings. The Sedge Warbler is still less particular as to site or materials, but generally uses some moss; its five or six eggs, moreover, are of an almost uniform yellowish brown, owing to the closeness of the spotting. They often have a black streak at the larger end, as in the similar eggs of the Yellow Wagtails. Breeding takes place early in May; while the Reed and Marsh Warblers do not begin, as a rule, till late in the month.

Most people have heard of the Grasshopper Warbler or "Fen Reeler" (*Locustella nævia*), so called from its notes, which resemble the sounds made by a cricket, a grasshopper, or the line running off a fisherman's reel. Certainly few would believe that they come from a bird's throat. Normally it only sings for a couple of months

Grasshopper Warbler

after its arrival in mid-April, yet it does not leave the country till September, though, being local and irregular in appearance, it may easily be overlooked. Moreover it is a skulking species, which loves sedgy or rushy flats or coarse grass on banks and at hedge-bottoms, while it does not despise clearings in woods and heathery moors. Imaginative writers even talk of it creeping like a mouse. Wicken Fen in Cambridgeshire and the Norfolk Broads are favourite resorts, but the discovery of the nest, always a matter of difficulty, there becomes almost an impossibility except to sedge-cutters. It is composed of grass and a quantity of moss, and contains five or six lovely white eggs with dull pink stippling. This Warbler has a characteristic habit of spreading its tail on being flushed from the nest, a fact which draws attention to an otherwise inconspicuous brown bird with darker streaks above and lighter tints below. It ranges in Britain as far north as Skye, and over the continent of Europe south of the Baltic, with south Norway and Finland; but eastward the limits are doubtful. The food consists mainly of aquatic insects and their larvæ.

Savi's Warbler (*L. luscinioides*) was only recognised as distinct by Savi in 1824, though Temminck had previously seen a Norfolk specimen and determined it as a Reed Warbler. To our country it was always an uncommon summer migrant, but it probably bred regularly in the Norfolk, Cambridgeshire, and Huntingdonshire fens, if not in Suffolk, until its disappearance in 1856. Southward from Holland, where it is now much rarer than formerly, it is found scattered over Europe, reaching to north-west Africa and west Turkestan. Savi's Warbler resembles its congener in general habits, but has a far harsher note; in colour it is of a plainer reddish brown. The nest is most peculiar, being composed of broad sedge leaves or of the grass *Glyceria aquatica*; the eggs are marked with purplish grey instead of pink.

Passeres

SUBFAMILY **Accentorinæ**, OR HEDGE-SPARROWS

We will now turn to the Hedge-sparrow and the Dipper, though the former has sometimes been placed nearer to the Thrushes, and the latter may possibly be akin to the Wren : the truth, which we should constantly bear in mind, being that no linear arrangement can ever accord exactly with nature, since a bird often exhibits striking affinity to species other than those to which it stands next in a list.

The Hedge-sparrow (*Accentor modularis*) is a resident or partially migratory species, the numbers of which are vastly increased by immigrants from the north in the cold season. It breeds far up our hills, though not in some of our bleakest islands, and occupies Europe from the northern limit of forest-growth to north Spain, as well as the Caucasus and Persia. Its shuffling gait on the ground, its weak flight, its mossy nest and five or six deep blue eggs are matters of ordinary knowledge, but its sweet note is sometimes mistaken for the less melodious strains of the Robin. The Hedge-sparrow breeds very early and rears more than one brood in the year, while it often acts as foster-parent to the Cuckoo. The nest may be found in low hedge-bushes, shrubs, heaps of brushwood, and the like, and almost invariably has a foundation of little dry twigs. In winter the bird will eat almost any scraps that are given to it, but the natural food is of insects, worms, spiders, and seeds. The colour is brown with blue-grey head and lower surface.

Family CINCLIDÆ, or Dippers

The Dipper (*Cinclus cinclus*) is one of many forms that occur in suitable spots throughout Europe and Asia, and extend to the Atlas mountains. By some our species

C. c. britannicus is considered distinct. The Water-Ousel, or Water-Crow, to use local names, is one of the most interesting of British birds from its striking appearance, its unusual habits, and its peculiar nest; it is a nice fat brown bird with a white chest and chestnut belly, which may be seen flitting from boulder to boulder on our rapidly running hill-streams, bobbing about on its stony perch and constantly diving into the

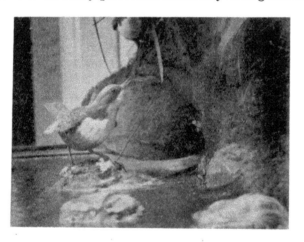

Dipper

water, where it uses its legs and wings below the surface. Naturally it is not suited by our southern and eastern counties, but it is found from Cornwall and Wales northward to the Orkneys, and also in Ireland, while its cheery song may be heard even in the severest weather in its upland haunts, which it seldom cares to leave. Its food consists of small mollusks, spiders, beetles, and insect-larvæ, and to some extent of fish-spawn, but its utility far outweighs its harmfulness.

Passeres

The nest is a big round ball of moss, or moss and grass, lined with leaves, generally of the oak or beech, and has a side entrance; it is built in a hole in a wall or bank, among protruding tree-roots on river sides or not uncommonly beneath a bridge or close to a water-wheel; in many cases it is stuck against a rock or large

Dipper's nest

boulder with the slightest possible support. The five or six somewhat pointed white eggs are laid from March onwards, and a third set has even been known to have been deposited in the same nest. The fledglings swim at once.

42   *Order I*

Family PANURIDÆ, or "Bearded Tits"

This remarkable family, of doubtful position, contains only one British member (*Panurus biarmicus*), a

Bearded Tit

slender and chiefly fawn-coloured bird with long black cheek-patches and an inordinately long tail, from which it is called " Reed-Pheasant " in Norfolk. The male has a grey head, the female lacks the black on the face. It is now confined in Britain to the Broads and Devonshire, but used to breed in other of the eastern and southern counties, and ranges abroad from temperate Europe to central Asia or even Manchuria, though limited by the fact that it requires large reed-beds to dwell in. These it seldom leaves even in the hardest weather, and there pairs or little parties may be seen by anyone who punts slowly past their haunts, flitting along the tops of the reeds, uttering their curious sharp note of "ping-ping," and soon dropping down in the vegetation. This note may even be heard in winter, and is only clearly audible on a calm day; in summer close observers may watch the hen building her nest of broad sedges or reeds and the cock bird supplying her with dry reed-flowers for the lining. The site chosen is always on the ground among the reeds, generally in open spots which hardly bear a man's weight, and eggs may be found from April to August; they are round and white with delicate blackish brown scrawls, and are about six in number. Many nests are cut over by the marshmen. The food consists chiefly of small mollusks and seeds of the reed, so that in almost every particular this species differs from the remainder of the Titmice, while its anatomy is certainly not that of a *Parus*. It has been introduced of recent years at Hornsea Mere in Yorkshire.

Family PARIDÆ, or Tits

Titmice are familiar winter-guests of the householder, and are resident in our islands, or migrate to a comparatively small extent; in fact different scientific names have been given to distinguish them from the various continental or Asiatic forms. All but one belong to the genus *Parus*, and in strictness should be called

Titmouses (German " meise "), though we hope that no one will follow such a practice. Our Long-tailed Tit (*Ægithalus caudatus roseus*) is a local race of *Æ. caudatus*

Long-tailed Tit

of Europe and Asia, and may also breed in France and the Pyrenees. It is distributed as far north in Scotland as trees occur, and over the rest of the kingdom is very seldom absent, though locally uncommon. The well-known nest is an oval ball of moss covered with lichens and warmly lined with feathers, while the beautiful fabric contains from seven to ten small white eggs,

Long-tailed Tit's nest

generally with indistinct pink spots. Forks of lichencovered birches are a favourite site in north Scotland; but thick hedges, gorse-bushes, small shrubs or the ivy on trees are commonly utilized; in various counties the bird is known as the "Bottle-tit" or "Feather-poke." After breeding these Tits keep in flocks, seeking for insects and larvæ, while they are more often seen than heard, for their notes are gentle; they fly pretty well, though unsteadily. The extremely long tail distinguishes them from other Tits, as does the black and white coloration, relieved by a rosy rump and belly best seen at close quarters.

The remaining species of Titmouse are all similar in their ways, flocking in winter, climbing about the trees, prying about the bark or leaves for insect-food, and laying from five to eight white eggs with bright red spots. But in certain points they differ. The three commonest are the Great, Coal, and Blue Tits, the Coal Tit usually predominating in pine-woods, and the others elsewhere. The Great Titmouse, or Oxeye (*Parus major*), which becomes rare in the north and north-west of Scotland, is sufficiently distinguished from its congeners by the broad black stripe down the yellowish under parts, though its black head and white cheeks are most conspicuous. In different forms it breeds south of the Arctic circle to the Mediterranean Islands, north-west Africa, Palestine, Persia, Burma, China, and Japan, while the British form has recently been separated from the others. It makes a nest of moss, surmounted by a felted mass of wool, fur and hair, in a hole in a tree or wall, when it does not choose a pump, a wooden letter-box, or other extraordinary situation. Its notes are comparatively harsh, and a rasping cry of two repeated notes which it often utters is commonly mistaken for that of the Chiffchaff; it also imitates the voices of other birds. The Great Tit is no doubt guilty of spoiling many fruit-buds, which may or may not contain harmful insects; it also eats peas, nuts, and seeds to some extent, and is always attracted by meat-bones, suet, cocoa-nuts and the like, hung up for it in

Coal Tit

winter. Occasionally it will even murder another bird, but insects are its staple food.

The Coal Titmouse (*P. ater britannicus*), a smaller species with no black stripe down the breast, has the nape-patch as well as the cheeks white; it does not breed in the Orkneys or Shetlands and abroad is represented by *P. ater*, with grey in place of olive back. Specimens, however, which many people think indistinguishable from the latter, are found breeding in Scotland, and the Irish bird, with a yellow rather than a whitish breast, has been lately separated; the truth being that it is difficult to draw the line between forms so closely related. The call-note of the Coal Tit is sharp and loud, so that it cannot easily be mistaken for those of its congeners; the nest resembles that of the Great Tit, but is usually placed in a hole in a wall or stump or in the ground, less commonly in a bank-side or the thatch of a shed. The bird is not so fond of an artificial nesting-box as the Great and Blue Tits are.

Until the last few years our somewhat browner backed Tits with merely the cheeks white were all united under the name of Marsh Titmouse, though not necessarily found near marshes. It has been ascertained, however, that two forms have been confounded under that title, the Marsh Tit with shiny black head in the adult (*P. palustris dresseri*) and the Willow Tit (*P. borealis kleinschmidti*) with dull black head. Even yet perfect unanimity has not been arrived at, and every reader of these pages must form his own opinion; but it appears that the birds breeding in Scotland and the Border counties of England should be referred to the Willow Tit, while further south both forms occur, with the Marsh Tit predominant in some or most of the districts, and especially in Kent. The latter seems to prefer woods, the former copses, hedgerows, and gardens. The Willow Tit cuts a neat round hole and lines the excavation with felted down, hair, and the like; the Marsh Tit does not usually cut its own hole and has a more decided substructure of moss. The Willow Tit's soft repeated cry resembles that of the Wryneck at a distance, the Marsh Tit's notes are merely somewhat sharper than those of the Blue or Coal Tits. Such at least is the writer's experience, and it is corroborated by that of others; yet further information is desirable, as the birds are only locally common, and it is no easy matter to find a nest in such a position as to enable the observer to lift the parent bird off the eggs and examine it. Irish birds also need further examination.

The Blue Titmouse (*P. cœruleus*) with its blue crown and nape, black throat and streak across the white cheeks, and yellow breast is a very familiar object in our gardens at any season; it does not extend in summer to the northerly isles of Scotland and similarly shuns the more Arctic parts of Europe. Otherwise it occupies the whole continent except Spain—in a form rather brighter than ours—and meets south of the Mediterranean and in Russia other allies which are sufficiently distinct to be characterized as species. Whether flitting along our hedgerows, hanging pendulous in search of insects on the trees, or engaged in building a nest in some hole of a tree or wall, the Blue Tit is always the same brave confiding little bird, which hisses violently at us when caught on its eggs, and resembles the Great Tit in its fondness for nest-boxes.

The Crested Tit (*P. cristatus*) is particularly interesting from the fact that it is confined in Britain to the ancient forests of the Spey and its tributaries. It has never been actually proved to breed elsewhere in Scotland, but a slightly different form occurs in many parts of the Continent. This local species is brown, with buffish white under parts, having the head and neck beautifully marked with black and white, the crest pronounced and erectile. In habits it is not unlike other Tits, but its note is rather loud and very characteristic. It inhabits Scotch fir woods, but does not invariably bore its nesting-holes in dead pines, as it occasionally chooses hardwood trees, or even ready-made holes in wooden or iron posts. The eggs are seldom more than six or seven and are particularly brightly marked.

Family REGULIDÆ, or Gold-crests

A very familiar bird is that smallest of British species the Gold-crested Wren (*Regulus regulus*), which ranges over Britain, as well as Europe, to the Caucasus and Asia Minor, within the limits of tree-growth—except Spain and Portugal. The continental form is hardly, if at all, distinguishable from that which occurs in this country, while it is one of the curious facts of nature that this delicate looking little creature migrates to and fro in immense flocks in autumn and spring, often for weeks together. We are never without the bird, which is often found in company with Tits. Its call-note is somewhat similar, but it has also a low song. It is fond of searching for its insect-food on fir trees, especially spruces, and on the latter it usually builds its beautiful little mossy nest lined with feathers, which is slung like a hammock under the tip of a bough and contains from five to ten white eggs ringed or covered with reddish buff markings. Alternative sites are the ivy on tree-trunks or small bushy conifers.

Constantly confounded with our common species is that rare immigrant the Fire-crested Wren (*R. ignicapillus*), which is also an olive-green bird with an orange crown margined on each side by a black band. Little reliance can be placed on the colour of the crown in males as a distinction, but the Fire-crested Wren is more sulphur-green on the neck and has a distinct black line through the eye. The females have lemon-yellow crests, the young have none. The last-named species does not range further north than the Baltic, but is the more common in some parts of southern Europe and of north-west Africa; it also breeds in Asia Minor.

Family SITTIDÆ, or Nuthatches

The Nuthatch (*Sitta cæsia*), notwithstanding its loud whistling spring notes, is a shy bird, which attracts little attention as it creeps quietly about the tree-trunks or flies heavily to a new hunting-ground of the same description in search of insect-food. The grey upper parts are not conspicuous and the buff breast is hardly more so. Nowhere really common, it is generally found in well-timbered districts and especially old parks, while it is only a casual visitor to Scotland and perhaps Ireland; the foreign range extends over Europe with the exception of Scandinavia and Russia, where a white-breasted form occurs, and Corsica, where there is a distinct species. As in the case of most of our residents it breeds early, choosing a hole or crevice in a tree— or more rarely in a bank—and lining it thickly with dry leaves or fir-scales, upon which lie the six or more white eggs with red and lilac spots, similar to, but larger than, those of the Great Tit. The nesting-hole is plastered up with mud, so as to leave only a circular entrance. This species is known as the Nut-jobber from its habit of hammering open with its powerful bill the nuts of which it is fond, and it is a pretty sight to watch it at work upon a hazel-nut stuck in the chink of a post. The winter notes are Tit-like, the spring notes shrill, the summer cries vary.

Family TROGLODYTIDÆ, or Wrens

Wrens have a wide range in the world, and unlike most Palæarctic families extend to temperate South

Wren

America; nevertheless we have only one British species (*Troglodytes troglodytes*), which is too common to need many details. Its simple but joyous song cheers us even in winter, and there are few road-sides where we do not often catch a sight of its restless little brown body and uptilted tail; it is found far up our hills and in our bleakest islands, while the numbers are increased in winter by migrants from abroad. The well-known big oval nest of moss, dry leaves, ferns, or

Wren's Nest

the like is often placed in most curious positions; the elongated white eggs are much less spotted than those of Tits, and are from six to ten in number. Our form of Wren occupies all Europe and does not reach west

Asia or north Africa, but variations in size and colour have caused examples from Iceland, the Færoes, Shetland, and St Kilda to be considered distinct subspecies.

Family CERTHIIDÆ, or Tree-creepers

The Tree-Creeper (*Certhia familiaris*) is found in Europe, northern, eastern and central Asia, and North

Tree-Creeper

America, and, as might be expected with so widespread a species, has been split into innumerable subspecies, with which we cannot concern ourselves here, though it should be noted that our race has been named *C. familiaris britannica*. Being a quiet little brown bird with whitish under parts it may easily escape notice, but it is never particularly common, though it is well known to country-folk as the Woodpecker, from its habit of creeping up the trunks and branches of trees, supported by its stiff-pointed tail feathers. The beak and claws are long and curved. Though it will eat seeds, the proper food consists of insects, in search of which the Creeper works spirally up a tree, finally flying off to begin at the base of another. It has a very low sweet song and a sibilant call-note. The nest has a foundation of small twigs below the main material of a little roots, grass, or moss, the interior being thickly lined with feathers, on which lie six or more very thin-shelled white eggs with red and lilac markings, like and yet unlike those of a Blue Tit. The nest should be looked for behind loose slabs of bark, but may be placed under eaves of sheds, in crevices of walls, or even in the foundations of large birds' nests, as is sometimes also the case with Tits.

Family MOTACILLIDÆ, or Wagtails and Pipits

A certain similarity may be observed between Wagtails and Pipits in their general habits, notes, and even nests and eggs, while systematists may now be said, on anatomical grounds, to be unanimous in combining them in one Family, though the Pipits have undoubted affinity with the Larks as well. The Wagtails are slim, lively, and confiding little creatures, with jerky undulating flight and a characteristic habit of keeping the hinder part of the body in constant motion when on the ground. It is very amusing to watch them on a grassy flat or a garden lawn; they make impetuous darts after insects for a yard or two, suddenly stop and almost fall forward on their heads to secure their fly or worm, and then make a fresh dart forward, or run with twinkling feet for a considerable distance. Ever and anon they fly up in the air after their prey, but this habit is more evident when they flit from stone to

Pied Wagtail

stone along a stream. Our Pipits have shorter tails than our Wagtails, and all are characterized by more or less white lateral tail-feathers. The Pied Wagtail, Water Wagtail, or Dishwasher (*Motacilla lugubris*) is our resident black and white species—resident, that is to say, in appearance, though large numbers pass to and from the Continent in autumn and spring, and even in Britain none are to be seen in the far north in winter. Abroad it breeds in Holland, Belgium, and north-west France, if not in south-west Norway; but in Europe generally, and perhaps even northern Africa, the representative form is *M. alba*, whether it be considered a distinct species or not. This form has, in the male, a light grey instead of black mantle, while even the females are much lighter. The young are less easily distinguished; and when, as is rarely the case, *M. alba* is reported as breeding in England, the male should always be carefully examined for fear of error.

The song is short and not very noticeable, but the sharp double call-note is familiar to all who live near water-sides in spring, for the birds must then always be near water, though the smallest runnels will often suffice. Later in the year they frequent our lawns and are common on the sea-shore. The nest is placed in a hole in a wall, bank, quarry, refuse-heap, or pollard willow, or among tree-roots projecting from a streamside; it is made of grass or roots and lined with hair and feathers. The half-dozen eggs are greyish white with small dark grey or blackish spots.

The beautiful Grey Wagtail (*M. boarula*) is to be seen throughout the year in the valleys of our hill-country, though very rarely at lower levels except towards winter; it is perpetually confounded with the Yellow Wagtail on account of its bright yellow breast, though its crown and back are grey and its throat black. This species ranges from south Sweden and mid-Russia to the Mediterranean, closely allied forms occurring in the Azores, the Canaries, and Asia. Holes in masonry, ledges of rocks, or projecting tree-roots by streams are the almost invariable sites for the nest; the structure is more mossy than in the Pied Wagtail, and the eggs are closely marked with brownish yellow. The true Yellow Wagtail (*M. raii*), which is greenish olive above and yellow below with browner wings and tail, is only a summer visitor to us, and frequents marshy flats, water-meadows and such places, though not uncommonly placing its nest in young corn or rye-grass. It is always built on the ground, in some depression of the soil, and is similar to that of the Pied Wagtail, though the eggs are even yellower in their markings than those of the Grey Wagtail, and have often the same black hair-line at the larger end. The Yellow Wagtail breeds locally in south Scotland, and also in western Holland and western France; it is, however, impossible in our limited space even to name the many allied forms that occur abroad. The best known in Britain is the Blue-headed Wagtail (*M. flava*), which breeds irregularly with us, chiefly in the south and east of England, and is common in most parts of the Continent; adults may be distinguished from those of *M. raii* by the blue-grey head and the white in place of yellowish stripe over the eye. In the adult male of the latter the crown is almost yellow.

Three Pipits breed with us, besides accidental visitors, and two are resident or partly migratory. Of these the best known is the Meadow-Pipit or Titlark (*Anthus pratensis*), abundant on our moors and by no means rare on rough ground at lower altitudes. It is found throughout Britain; from Iceland to west Siberia; and thence to the Pyrenees, north Italy, and

Palestine; south of which limits it is practically a bird of passage, for *A. bertheloti* of Madeira and the Canaries is considered a distinct species. As the Titlark is a plain streaky brown bird with white breast striped with the same colour, it would not be specially conspicuous, were it not for its habit of soaring a little way up in the air to utter its shrill song, and flying restlessly round an intruder while giving vent to the sharp alarm notes, which account for one of its names; this generally takes place near the nest—a plain cup of bents placed in some depression of a rough grass field, on a bank, or among heather—which contains about five white eggs very thickly marked with brown. The Cuckoo is often reared by this species, and almost invariably so on the moors, which afford the regular food of insects, worms, small mollusks, and seeds in abundance.

The less demonstrative and more local Tree-Pipit (*A. trivialis*) differs little in appearance, but has a much shorter and more curved hind-toe. It only visits us between April and September, and is not found in the northern islands of Scotland or in Ireland. Except for Iceland and the Færoes, the foreign range is much as in the last species, though a separable race extends further eastward, to Japan and China. In habits, however, it is absolutely different, for it frequents open copses or the outskirts and rides of woods, where it pours out its sweet notes while sitting on the tree-tops or while soaring to a considerable height in the air above them. Away from these quarters it is seldom seen, and there it builds a similar nest to its congeners, but lays very remarkable eggs. No British bird, except the Guillemot, shews such a range of coloration in the markings, which may be purple, red, rich brown, or almost black, while the pinkish or greenish ground-colour varies in accordance. In Scotland the Tree-Pipit is often called "Wood-Lark," a very natural mistake where that species does not occur.

The Rock-Pipit (*A. petrosus*) is larger than the Meadow-Pipit, and rather more olive, while the outer tail-feathers are marked with smoke-colour rather than white. It is entirely confined to our rocky shores, but ranges abroad from Norway to north and west France; elsewhere matters are complicated by the occurrence of a form with a reddish breast. Since Meadow-Pipits also breed on our cliffs, observers must be careful in their identification, for the habits as well as the plumage are similar, and both birds feed on the shore after the breeding season. The Rock-Pipit's eggs are merely a little larger.

Family ORIOLIDÆ, or Orioles

The Golden Oriole (*Oriolus oriolus*) might be relegated to our list of irregular migrants, were it not for the fact that it now breeds in Kent and occasionally in our eastern and southern counties. All should therefore be on the look-out for a beautiful bird of the size of a Thrush, golden in colour with black wings and tail, and use every means to preserve it, if seen, or permit it to breed in safety. The nest is a sort of cradle of grass, wool and bast, slung in the fork of a branch; the large eggs are white with round purplish black spots. The Golden Oriole has a swift but heavy flight, a lovely flute-like song and a harsh call, but it eats too much ripe fruit to be popular abroad, though insects form part of its diet; it often frequents town gardens in Europe, south-west Asia, and north-west Africa, and is one of the brilliant members of a large family extending over a great part of the globe. The female shews little yellow.

Family LANIIDÆ, or Shrikes

The only member of this family that breeds regularly in our islands is the Red-backed Shrike or Butcher-

Red-backed Shrikes

bird (*Lanius collurio*), which gradually decreases in numbers as far north as southern Scotland and is only a straggler to the north of that country or to Ireland. Arriving early in May it is a conspicuous object as it sits on the tops of bushes, telegraph-wires, and the like, while its big nest of grass and moss is hardly less conspicuous than the bird itself, in the solitary hawthorn bushes which it selects by preference. It may, however, be placed in a thick hedge or shrub, or on a low branch of a tree, while it is always lined with wool and hair, and contains some five eggs varying in tint from reddish to greenish white, with fine blotches or spots of the corresponding colour, not uncommonly collected into a zone. This Shrike is very wary, but fairly bold in the breeding season, after which it is little seen till its departure in August: it feeds on small mammals and birds, beetles, bees, and other large insects, and occasionally keeps a small stock impaled on thorns near the nest. These "larders," however, are not so common as has been supposed. For short distances the flight is strong, while the bird has a harsh and somewhat Chat-like note, as well as a slight song. The foreign range extends over northern Europe and Asia to Transcaspia and north Persia, and over southern Europe except the Iberian Peninsula. The male is chestnut-brown above with grey head, black and white tail, and black face, and pinkish buff below; the female is red-brown above, and whitish below with crescentic markings recalling those of some hawks.

The Woodchat Shrike (*Lanius senator*), distinguished by its chestnut head and white wing-bar, is a common continental species which visits us at very irregular intervals, but must be mentioned here as having possibly bred twice in the Isle of Wight. It is therefore one of the birds for which a watch should be kept in the south.

Besides these we have two other members of the family that occur in Britain, the Great Grey and the

Passeres

Lesser Grey Shrikes. The latter (*L. minor*) is only a straggler from south and central Europe, but the former (*L. excubitor*) is a pretty regular immigrant in the cold season, especially to our eastern coasts, though it has never been known to breed with us. It comes from northern and central Europe, but the range can hardly be defined here, on account of the various species or races that have been described from Europe, Asia, north Africa, and North America. It is very doubtful whether the form that occasionally visits us, with one white wing-bar (*L. major*), is separable from the typical form with two bars. The colour is grey above and white below, with black cheeks, wing- and tail-feathers. The nest is larger than in the case of the Red-backed Shrike, and the eggs are greenish white with olive markings.

Family AMPELIDÆ, or Waxwings

Perhaps the most beautiful of our constant but irregular visitors is the Waxwing (*Ampelis garrulus*), a bird of many colours. It is mainly fawn-brown, with more chestnut head, crest, and lower tail-coverts, black cheeks and throat, and with yellow and black on the wings and tail. Added to this the tips of the shafts of the tail-feathers and of many of the secondaries are wax-like and scarlet. In some years few visit us, in some great numbers arrive, chiefly on our northern and western coasts, and may be seen satisfying their hunger on the berries untouched by the other birds; they generally come in our hardest winters, driven from their summer haunts in Arctic Europe and Asia, where they are very changeable in their breeding quarters, which extend eastwards at least to Alaska and the Rocky Mountains. The nest was unknown until Wolley procured

64   *Order I*

it in Lapland in 1856, and is an unusual structure like a platform of twigs, surmounted by a large open cup of lichens and grass; the eggs also are of a peculiar grey-blue tint with roundish blotches and streaks of blackish brown and lilac. The Waxwing's low continuous note is not much heard, and the bird is shy in

Waxwing

summer, when its food consists mainly of insects. Its flight is strong and often high.

Family MUSCICAPIDÆ, or Flycatchers

We next come to a much more modest species, the plain brown Spotted Flycatcher (*Muscicapa grisola*), so called from its streaked breast; it is very late in arriving from the south, and only remains with us from May to September, while it ranges over the Palæarctic region eastwards to Lake Baikal and Dauria, except the extreme north. The low but pleasing song and the sharper call-note are familiar to most of us, and may even be heard in London; but this confiding bird is best known from its habit of sitting on a post,

Spotted Flycatcher

stake, or the outer branch of a tree, whence it is constantly darting out after the insects which it captures on the wing. Thus it is often termed Post-bird, and another name is Beam-bird, from its fancy for a beam on which to place its nest. More common sites, however, are creepers on buildings, hollows in broken or pollard trees, and holes in walls, besides many curious positions.

The five pretty eggs are spotted with rufous on a greenish-white or even a green ground, and lie in a mossy nest, lined with warm materials, and often adorned with lichen. This species does not breed in the Hebrides, Orkneys, or Shetland.

That much rarer bird the Pied Flycatcher (*M. atricapilla*) has a more restricted range abroad, where it only extends southward in its various forms to north Africa and eastward to Persia and Palestine, while, being a particularly arboreal species, it is decidedly local. In Britain, where it remains from May to August, it breeds chiefly in the west, from south Wales to Cumberland, irregularly in Scotland, but not in Ireland. Artificial boxes in our shrubberies have proved a great attraction to this bird, a somewhat curious fact, as it naturally frequents water and seldom leaves the sides of open shady streams, where it makes a pretty picture as it flits, Warbler-like, from oak to alder or ash, often uttering its sweet little song, and the male in particular exhibiting his bright black and white colours in contrast to the brown female relieved by dusky white. The insect-food is very commonly taken on the wing and conveyed straight to the brooding hen. The nest is composed entirely of roots and grass lined with hair, and thus can easily be distinguished from that of the Redstart, which has precisely similar pale blue eggs. It is always in a hole, and generally in a tree.

Family HIRUNDINIDÆ, or Swallows

Three members of this family are common and well known in Britain, the Swallow (*Hirundo rustica*) with long streamers or outer tail-feathers, chestnut throat and buff under parts separated by a metallic blue chest-band, the House-martin (*Delichon urbica*) with shortly forked tail, conspicuous white rump and lower surface, and the Sand-martin (*Riparia riparia*), which is not blue-black above like the others, but is brown with a mottled band of the same colour on the white breast. It has, moreover, a tuft of feathers above the hind-toe. All three have very short beaks with a wide gape. It will be seen below that the Swift is not a Swallow, but a "Picarian" form allied to the

Swallow's nest

Humming-birds. The nesting habits are as distinct as the coloration, for both the Martins breed in colonies, while the Swallow does not. It builds an open cup with pellets of mud, and lines it warmly with straw and feathers to hold the five or six white eggs with brown and greyish markings; the House-martin sticks a half-cup of the same substance, with an aperture at the top of one side, under eaves, in window corners, under rock-shelves or mouldings at the top of bridges, adds a bedding of straw, chaff, or softer materials, and lays four or five pure white eggs; the Sand-martin makes a grass or straw nest thickly lined with feathers at the end of a tunnel, which it bores in a bank or even a big heap of sawdust, and also lays white eggs. Sometimes it uses a hole in a wall.

All these species are migrants, and are with us from about the end of March to November at the latest,

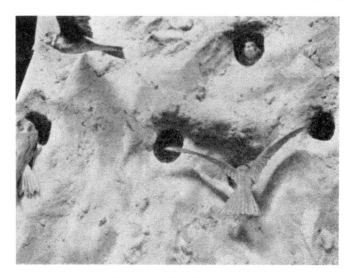

Sand-martins' burrows

but the Sand-martin is generally observed first and departs a little earlier; moreover, it breeds in the Nearctic region, in north-west India, and on the Nile, as well as in the Palæarctic region, while the other two species are confined to the latter. So too it is found in the Neotropical region as well as in the south of the Old World in winter, while the typical forms of the Swallow and the House-martin do not migrate to

America, though they go further south in Africa and reach the Cape of Good Hope. When about to leave for their winter-quarters, the members of this family collect into flocks, especially in the case of the Swallows, which just before crossing the sea may be seen massed on roofs, telegraph-wires, and so forth, or even crowded together in huge quantities on the roads. The twittering notes and the flight are well known to everyone; the food consists of insects taken in the air or from the surface of the water. The birds always seem too busy to be shy, and are too quick and irregular in their movements ever to be in much danger from shooters; occasionally, however, they take it into their heads to attack a man, swooping down with much noise and brushing him with their wings, for no apparent object. Swallows commonly perch on trees, Martins rarely. All may have two or even three broods, especially when their first nests are usurped by Sparrows.

Family FRINGILLIDÆ, or Finches and Buntings

The members of this very large Family are for the most part stout-looking birds with strong bills, which feed mainly upon seeds and fruits, though the diet of the young consists also of insects and their larvæ. They vary considerably in size and coloration, as may be seen from the Sparrow and Linnet; the Hawfinch and Crossbill; the Goldfinch, Bullfinch, and Chaffinch; the Common, Yellow, and Snow Buntings. The subfamily *Fringillinæ* contains the true Finches, where the mandibles fit closely together, the subfamily *Emberizinæ* the Buntings, where the bill, when shut, shews a distinct gap. This seems a small point, but the subdivision is in agreement with the appearance and habits of the birds.

SUBFAMILY **Fringillinæ**, OR FINCHES

The Greenfinch or Green "Linnet" (*Chloris chloris*), one of our most familiar residents, breeds throughout Britain and Europe generally, south of the Shetlands and the more Arctic districts; it has also been reported from western Asia and north-west Africa, but southward and eastward closely allied forms forbid exact limits to be defined. The olive-green colour, relieved by a little bright yellow and blackish brown, is best seen as the family parties flit before us along the hedgerows in summer, while in the breeding season the monotonous droning call-note is most usually heard as the bird sits unseen in the dense foliage. The flight is strong, and large numbers arrive from abroad towards winter; the song is feeble, and the flocks on the stubbles are pretty quiet. The nest, placed in a tree-top at some height from the ground, in a shrub or a hedge, is composed of roots, moss and wool, and lined with wool, hair and feathers; the four to six eggs are greenish or reddish white with red-brown spots, as in a true Linnet.

The Hawfinch (*Coccothraustes coccothraustes*) is a most interesting species of peculiarly heavy build, with a comparatively short tail and immense beak; it used to be a rare resident in England, and is still extraordinarily erratic, but has certainly spread northwards of late years to Dumfriesshire and Fifeshire, besides becoming locally common in the south. It breeds throughout the Palæarctic region, except the more northern parts, if we do not separate the eastern Asiatic and other supposed western forms. Its food consists of seeds and fruits, such as haws, beech-mast, kernels of cherry-stones, and peas, with caterpillars for the young, so that it finds its strong beak of the greatest use. Rows of peas are often nearly stripped when the young have left the nest. This structure, composed of roots and grass, often with an admixture of lichens and a base of twigs, is placed in the fork of a large hawthorn or fruit-tree, if not towards the top of a pollard or on some horizontal branch, and contains five or six curious bluish or greenish eggs with fine olive and greyish markings, which may be either spots or scrawls. Breeding takes place in May, when the bird is remarkably shy, as indeed is its usual habit. The flight is laboured. The characteristic whistle, however, soon draws attention to it; the song, on the other hand, is inconsiderable. After the nesting season the Hawfinch often wanders about the country, and it is then that we not uncommonly hear reports of a big fawn-coloured bird with a black throat, grey neck, and blue and white on the wing having been seen in a garden.

The Goldfinch (*Carduelis carduelis britannica*), one of the few species that live not unhappily in a cage, though a delicate-looking bird of the Canary type, is also partly fawn-coloured, but is beautifully marked with red, yellow, black and white. It is difficult to form a clear idea of its numbers in England as, after a serious decrease in the nineteenth century, it seems to be undoubtedly increasing again; but nests might now be found in any county, while such has been the case across the Scottish Border even as far north as Perthshire and Skye. Many migrants pass to and from the Continent at the usual seasons, and scientific names have been given to several more or less distinct forms; it is impossible, therefore, to lay down any exact limits for each, but they occur throughout Europe, north

Africa, and western Asia. The bird's fondness for the seeds of thistles, groundsel, and other composite plants is unfortunately well understood by bird-catchers, who take many individuals in their clap-nets when and where they have no business to do so; but the sweet song is little appreciated save in captivity, though country-folk are generally able to recognise the clear call-note. The Goldfinch often escapes observation as it passes lightly above our heads, for its bright colours shew up badly on the wing; while the nest, a pretty little cup of moss and wool with the softest of linings, is usually well hidden at the top of a fruit or other tree, though occasionally placed in a hedge or shrub. The four or five eggs are very pale bluish white with purplish or blackish spots.

The Siskin (*Spinus spinus*) is comparatively rare, and in Britain only breeds regularly in eastern and south-western Scotland and in Ireland, though occasionally in various English and Welsh counties where spruces and larches abound. From autumn to spring, however, flocks are seen in many other districts, feeding on the seeds of the alders and ragwort or flitting about the country-side; yet the bird is local with us and also abroad, where it inhabits northern and central Europe and Asia as far as Japan. Perhaps it is best known in cages, for it is a favourite on account of its sweet, if not varied, song and the bright colours of the male, which is greenish olive with yellow breast and rump, black and yellow wing and tail markings; and also has a black crown and throat. The female shews much less yellow and no black above and is streaked, especially on the whitish under parts, with brown. The flight, the nest and eggs are much as in the Goldfinch, but the nest is usually high up in a silver fir or larch, and placed on a horizontal branch ; sometimes it is much nearer to the ground or even in a bush. The substructure is commonly of little twigs and the eggs are bluish in tint with dark spots.

We need not enter into any details about the colour or habits of the House Sparrow (*Passer domesticus*), but may observe that the harm it does by interfering with the nests of the insectivorous Martin and devouring grain is to a great extent counterbalanced by its destruction of insects and their larvæ. South of the Arctic Circle the Sparrow is said to range to Spain and Austria and also to Irkutsk and Dauria in Asia, but it is only found near the dwellings of man, and is represented in parts of south Europe by the possibly distinct Italian and Spanish Sparrows.

A second resident species, the Tree Sparrow (*P. montanus*), needs careful observation to distinguish it from the last-named, as, roughly speaking, it occupies the same localities both at home and abroad, though it is curiously local, while it extends to the Mediterranean, Siberia, China, Japan, and possibly northern Africa. The crown and nape are rich brown, the white cheek shews a black patch, and the wing two white bars, while in the cock Sparrow the crown is grey, there is no cheek patch, and the wing has only one bar. Moreover the sexes are alike in the Tree Sparrow, whereas everyone is familiar with the dowdy brown hen of the other species. In both the cocks have black throats. Again the large untidy nest of the House Sparrow is placed in spouts, gutters, holes in masonry, banks or trees, and also in hawthorns or the like in the open; that of the Tree Sparrow is invariably in a hole, whether it be in a wall, railway cutting, pollard willow or whatnot, and is built with much less straw and feathers, while the eggs are decidedly smaller and more delicate than those of its congener, and generally marked with deep brown instead of black and grey. In fact, just as the House Sparrow's eggs are rather like those of the Pied Wagtail, those of the Tree Sparrow are generally like those of the Meadow Pipit. The chirping and chattering notes of the two species are also easily distinguished by those well acquainted with them.

The Chaffinch (*Fringilla cœlebs*) is resident with us, though its numbers are greatly augmented in autumn by flocks which come from the north, for the bird ranges over Europe from the Arctic Circle southward and also across west Siberia. It is too well known to need description, but in north Africa and the Atlantic Islands the allied species are most interesting and practically to be considered distinct. After the breeding season the males keep by themselves to a considerable extent till spring, and thus the Chaffinch gains the name of "*cœlebs*," or bachelor. It is only too tame and familiar in most places, and is very fond of scratching up and eating newly sown seeds of cabbage, turnip or radish; yet both cock and hen are general favourites, and their call-note of "spink-spink" is always pleasant to the ear. The song varies considerably, but is always of the same stamp; the nest of moss and wool, studded outside with lichens, is one of our prettiest pieces of bird-architecture; and the eggs, which are spotted with red-brown, have a curious dull greenish ground-colour, or are rarely quite blue.

The Brambling or Mountain-Finch (*F. montifringilla*) arrives on our eastern coasts about October, and only remains with us till March, for supposed instances of its breeding in Britain are erroneous. Huge flocks often visit us, but are much rarer in the west and in Ireland; abroad it spends the summer in Subarctic Europe and Asia, where it builds a rougher nest than the Chaffinch on the trees or bushes that are available, and lays very similar eggs. It is extremely fond of beech-mast, and is therefore commonly met with in winter on the ground under beech trees, but it also eats seeds in general as well as insects. The summer habits are a little different from those of its congener, as might be imagined in a far northern climate, while the song, if it should so be called, is harsh and jarring. The black upper parts, and white for green rump distinguish the male from the cock Chaffinch. The female is chiefly brown above.

That well-known cage-bird the Linnet (*Acanthis cannabina*) is a resident in Britain in the usual sense of the word; that is, the flocks move southwards within the kingdom in winter and are joined by large numbers from the Continent. Abroad it ranges in its various forms over the Palæarctic region to Kashmir. Bird-catchers know it as the Grey, Brown, or Red Linnet, the reason for the various names being that breeding males have the brown plumage decorated with red on the forehead, crown, and breast, while hens and individuals not in perfect plumage exhibit no red, which is, moreover, generally lost in captivity. Restless at the nest, and shy at all times, Linnets are usually seen flocking to the stubbles to feed, flitting about gorse-covers, or passing overhead with a somewhat jerky flight as they utter reiterated twittering notes; in a cage the song seems to become more full and to consist partly of imitations. The food is of grain and seeds and partly of fruit. The nest of roots, lined with wool and hair, is placed in low hedges, bushes, or even in rough herbage; gorse-covers are favourite spots, and large colonies have been found in thickets of privet. The five or six eggs are greenish or bluish white with rufous spots, large specimens resembling those of the Greenfinch.

The Twite or Mountain-Linnet (*A. flavirostris*) is often mistaken for the Linnet proper, but is readily distinguished at close quarters by the very short beak and the rose-red rump of the male; it represents its congener on our higher hills and moorlands from Devon, central and northern England, to Scotland and Ireland, being specially abundant in the northern isles. The "Heather Lintie," as it is called in Scotland, only breeds abroad from Scandinavia to Finland, if we separate a paler Asiatic form. Its habits in general are those of the Linnet; but the nest is either in the heather or on some grassy slope, often near the sea, while the eggs are distinctly bluer and rounder than those of its congener.

The Redpolls, as a genus, have given a considerable amount of trouble to ornithologists, who estimate quite differently the value of the various forms—which are worthy of specific, which of subspecific rank. Apart from accidental occurrences, however, we are only here concerned with two of these, the Mealy Redpoll and the Lesser Redpoll, and they are sufficiently distinct. The Lesser Redpoll (*A. cabaret*) is best known in the cold season, when flocks, great or small, are seen feeding on the seeds of birches, alders, or conifers, that form the chief article of their diet; they generally fly high overhead, uttering their twittering cries, which resemble, but can readily be distinguished from, those of the Linnet. They change their feeding-grounds continually and are strong on the wing, yet they can hardly be called shy, and are fairly tame in the breeding season. The male of this smallest of our Finches is a pretty little brown bird with black throat, red crown and breast, and pinkish rump; the female has no red save on the head, and even this is lacking in the young. The nest, placed in a tall hedge, a shrub, or a low growth of poplar, alder, willow or hazel, is even prettier than the builder, being composed of grass, moss, and wool, thickly lined with willow or poplar down, where such is obtainable; the half-dozen eggs are of a darker greenish blue than in the Linnets, and have rufous blotches. This species nests locally from the Orkneys and Inner Hebrides to the south of England and Ireland, while abroad it is by some considered to range over western and central Europe. But here again the question of forms makes a definite statement dangerous. It certainly breeds in the Alps.

The Mealy Redpoll proper (*A. linaria*) migrates to our northern and north-eastern coasts from the far north after breeding; it is most common in hard winters, and is little seen south of Yorkshire, though when it does come it generally comes in flocks; in the west and in Ireland it is rare. It is a larger and much lighter bird than the last species, being streaked with white above; in habits, however, it hardly differs, allowing for any changes due to a residence in the extreme north of Europe, Asia, and America; down is there less easy to procure for the nest, which is a little larger, as are the eggs.

The Bullfinch (*Pyrrhula pyrrhula*) is so well known

in cages that it would hardly be necessary to call attention to its coloration, were it not that the conspicuous black head, grey back, white rump and red breast easily enable the male to be identified out of

Bullfinch

doors. The female is browner above and vinous-brown below. It is a shy bird, a frequenter of woods and thickets in the breeding season, and more or less retiring at any time. South of the latitude of Skye it is generally common in suitable localities, though often overlooked;

while many gardeners shoot it at sight for the damage it does in spring to the buds of fruit-trees, possibly in search of insects. These with their larvæ, seeds and berries constitute the food. The flight is rather heavy, the notes low and mournful, but they develop into a fine piping song in captivity, for the Bullfinch is an apt pupil when taught. The nest, built in a bush, creeper, or thicket, is made of roots lined with hair, and contains four or five bright greenish blue eggs with purple spots; it has invariably a foundation of dry twigs. If we separate the northern European and western Siberian bird from ours it will stand as *P. pyrrhula* proper, and the British as a subspecies *pileata*.

The Crossbill (*Loxia curvirostra*) has been much in evidence since 1909, when larger flocks of the Continental race than usual appeared in June and July, and remained to breed in 1910. They were not confined to any single district, but were most abundant in the conifer woods near Thetford in Norfolk—where some pairs are still nesting—in 1914. The local keepers assert that a few birds have always bred there, and exceptional instances for other counties had been on record for many years back; but the Crossbill used only to be known as a regular inhabitant with us of the fine old Scotch-fir woods of the shires from Aberdeen and Inverness northwards, where the local race is by some considered to differ from the Continental and is named *Loxia curvirostra scotica*. This species has now nested for a considerable number of years in Ireland. It builds a compact or sometimes careless structure of twigs, grass and moss, not uncommonly decorated with lichens, and lays four or five eggs a little larger than, but similar to, those of the Greenfinch. Scotch-firs are preferred, but larches are also utilized, and the nest is more often on a horizontal branch than near the trunk of the tree. The male is very conspicuous when he sits in the morning on some tree-top, pouring forth his lively song and shewing off his crimson colours; only birds in perfect plumage, however, are so bright, the remainder being orange and the females greenish, while all have browner wings and tail. The young are much greyer. The Crossbill's presence is generally made evident by the number of fir-cones with their seeds extracted which lie below the trees, but insects of different sorts, caterpillars and fruit give variety to the diet. As cones become scarce the flocks move to new places, and their low sibilant notes are heard no more; the flight is strong, as might be expected from the size of the bird. This species, in slightly varying forms, breeds in the conifer-districts of both the Palæarctic and the Nearctic regions, a specially stout-billed race having been denominated the Parrot Crossbill. The characteristic crossing of the tips of the mandibles is not peculiar to the Crossbills; it has, for instance, occurred exceptionally in Redpolls.

SUBFAMILY **Emberizinæ**, OR BUNTINGS

The Corn-Bunting (*Emberiza calandra*), a heavy plain brown bird with whitish under surface streaked with the same colour, is with us at all times of year, though a certain amount of migration is known to take place.

It breeds in various places throughout the Palæarctic region save in the far north and east, and is abundant, though local, in Britain. It prefers uncultivated districts and those without woods, not ascending to any great altitude, and being nowhere more common than in Orkney and Shetland. Its heavy flight and droning note distinguish it from all our other Buntings, and it is always in evidence during summer on its favourite perches, which are usually a bush-top or a telegraph-wire. After breeding, it is often seen in flocks, and lives almost entirely on grain. The nest of grass and hair is placed among coarse herbage or in young corn, and the large eggs, which are laid in May or later, have

Corn-Bunting

a yellowish or purplish ground-colour, finely blotched and streaked with browns and lilacs.

A brilliant yellow head, equally bright under parts, and a red-brown rump, hardly marred by brown streaks, are the well-known characteristics of the male Yellow Bunting or Yellowhammer (*Emberiza citrinella*), while the song of "a little bit of bread and no cheese" and the eggs beautifully spotted and scrawled with purplish or reddish brown are familiar to us all. In most other respects it resembles the Corn-Bunting, that is in flight, food, choice of perch, nest, and sociability. It breeds, however, several times in a season and often builds its nest in young trees or bushes. Northwards it is much less common than its congener and abroad hardly extends southwards beyond the Pyrenees and the Alps. The female is much browner.

The rarer and more retiring Cirl Bunting (*Emberiza cirlus*) of central and southern Europe and north-west Africa is also a resident in southern England, but has not been proved to breed north of Yorkshire and rarely does so in north Wales. It may almost be called abundant, however, in parts of Hampshire and Devonshire. If a clear view is obtained, the male is easily distinguishable from the Yellowhammer by its black throat, lores and ear-coverts, and much less yellow head and lower parts. The female is almost brown above with a buff instead of a black throat. The song is like the first part of that of the Yellowhammer, but has not the concluding phrase; the flight and nest are similar, but the eggs have a somewhat blue ground-colour and more spots than scrawls. The site chosen is usually in some dry sunny locality near houses, especially on chalk downs or west-country "combes," and the actual place most commonly a low bush.

It is rather unfortunate that we have been obliged to transfer the popular name of Black-headed Bunting from *Emberiza schœniclus* to a foreign species, nor does the decision seem absolutely necessary ; but it is useless now to complain, and fortunately we have a good alternative in "Reed Bunting." The black head and breast, white collar and belly, and mainly red-brown upper parts make the male a very conspicuous object as he sits on the top of some bush or clings to the highest reeds, whence he flits along before an intruder to another similar spot. He is by no means shy and constantly utters a sharp call-note, besides which he has a clear drawling song of the type usual in Buntings. The hen is brown with reddish head and black streaks on the throat. In winter the birds sometimes gather in flocks, but generally frequent the fields and especially the ditches; for there seems to be little migration to or from the Continent in the case of our home-bred birds, though occasionally numbers cross to us from abroad. There the range extends over all the Palæarctic region, except north Africa, if we do not separate smaller or much larger billed forms; in Britain this species has not been found breeding in Shetland. The food includes insects from the marshes, small crustaceans and mollusks, grain and seeds of various plants. The nest, of grass and moss or dead marsh-herbage, is placed in a low shrub, or in vegetation just clear of the ground, but always in damp spots; it is lined with hair or dry reed-flowers and contains about five brownish white or distinctly green eggs with purplish brown markings, which may include a few scrawls.

The Lapland Bunting (*Calcarius lapponicus*) was, if not overlooked, a decidedly rare winter migrant to Britain until 1892, when a large irruption took place, followed by others in each subsequent year. Its habits are more those of a Pipit than a Bunting, and with us it generally keeps to the shores and coast-lands. The black crown and face of the male are in striking contrast with the chestnut nape and white sides of the neck; otherwise the bird is chiefly brown, while the female is much plainer brown. As its breeding haunts, though circumpolar, are in the Arctic and Subarctic countries, the eggs were for long rare in collections; they are richly spotted with brown and are laid in a grassy nest warmly lined with feathers, which is placed on a dry spot in a marsh—commonly to the north of the tree limit—in a willow swamp or similar place. The song is said to resemble that of the Linnet.

The Snow Bunting (*Plectrophenax nivalis*), also an inhabitant of the more Arctic regions of both worlds, as well as Iceland and the Færoes, is of particular interest to British ornithologists, for not only do large numbers visit us between October and April, but a good many pairs are now known to breed at the tops of the loftiest Scottish mountains. There the nest, of such substances as can be procured, is placed deep among the boulders of the "screes," but in the north of Europe it is often more exposed and almost at sea level. The five roundish white eggs are prettily marked with rust-colour, brown and lilac. The song is more melodious than in Buntings generally; the flight is strong; the food consists of insects in summer, while in winter the birds frequent our sea-side dunes, fields, and stackyards for the seeds and grain to be found there. The cock is a beautiful white bird with black on the mantle, tail and wings, which becomes chestnut in autumn; he is whiter again in winter, but the hen and the young are always much duller, the former being greyer with blackish head.

Family STURNIDÆ, or Starlings

It is difficult to believe that the Starling (*Sturnus vulgaris*), which is now so abundant from Shetland to

Starling

Cornwall, was hardly known north of the Scottish Border at the beginning of last century. The increase of late years has been enormous, and it is now no uncommon thing to discover a winter " roost " of many thousands, which literally break down reeds, shrubs or even small trees by their combined weight. Abroad our bird is sufficiently common throughout most of Europe except the far north, Spain, Portugal, Sardinia, Corsica, Sicily, and the Færoes; in Asia a form occurs in Siberia and from Asia Minor to north India. The question of its range is complicated by the occurrence of an unspotted species in south Europe. The buff markings on the black plumage with its green and purple sheen are only characteristic of the adult Starling, for the young are plain brown with duller yellow bill. In this state they are often seen in autumn flocking to our shores, where migrants also arrive from abroad; but the birds are more familiar to us on our houses and in our woods, where they make a very untidy straw nest in almost any sort of hole and lay some six pale blue eggs. Occasionally heaps of stones on roadsides are made use of. Starlings are decidedly beneficial to the farmer, as they eat huge quantities of slugs, worms, insects and their larvæ; the only harm they do is to ornamental berries or fruit, if we except the few small birds they are occasionally known to kill. Wary but by no means shy, their funny ways and queer little hurried runs when feeding on the ground are known to every observer, but the varied shrill notes of these admirable mimics often delude him as to their author. The flight is strong, but somewhat spasmodic. Our breeding stock is partially migratory.

The Rose-coloured Starling (*Pastor roseus*) with its beautiful rose-pink body, black wings, tail, neck, and crested head, is an irregular visitor to Britain, generally between spring and autumn, when it attaches itself to companies of Starlings. It is a south-eastern European and west Asiatic species, occasionally flocking to Italy and Hungary, which breeds in large colonies, that constantly change their quarters, perhaps according to the food-supply. A favourite article of diet is the migratory locust, but other insects, berries, fruit and grain are also eaten. The nests of grass and feathers are placed in holes in old ruins, cliffs, and banks; the eggs are bluish white. It is a comparatively shy bird, with the general habits of a Starling, and a harsh chattering cry.

Family CORVIDÆ, or the Crow Tribe

The glossy black Chough (*Pyrrhocorax pyrrhocorax*), with its brilliant red bill and legs, was evidently a much commoner bird in Britain of old than at the present day, when it barely holds its own on the cliffs of the west and in Ireland. On the east the last nest appears to have been recorded about the middle of the nineteenth century, at St Abb's Head in Berwickshire. It rarely breeds inland in our country, but it does so in many parts of the high mountains of western and southern Europe, Asia, Abyssinia and north-west Africa. It also inhabits the islands of the Mediterranean, the Canaries and the coasts of west Europe, with the exception of Scandinavia. This shy bird is non-migratory and may therefore be seen at any season actively hunting for food on the ground, after the manner of the Starling, but it has no special fancy for fruits. Occasionally it utters notes resembling "chough-chough," but the usual cry is clear and ringing; the flight may be prolonged, but generally consists of circling movements varied by tumbling. The nest, placed in some hole in a steep slope, a cave or cliff-face, and often most difficult of access, is made of twigs, heather and the like, with wool and hair for lining; the four or five

Jay

eggs are yellowish or greenish white with markings of grey, brown, and lilac.

Passeres

A wary and subtle bird is the Jay (*Garrulus glandarius*), as much admired for its beauty as hated by gamekeepers for its destructive tendencies. Fawn-colour, black and white are mingled in its plumage in due proportion, while a large crest and a mottled blue wing-patch enhance its attractiveness. The food consists of worms and insects, acorns, nuts, and other fruits, and unfortunately also of the eggs and young of birds; the flight is heavy and the reiterated notes harsh and screaming. The nest, sometimes built in the fork of a tree, but ordinarily in thick bushy copsewood, is of twigs, grass and roots, while the four or five greenish eggs are closely freckled with olive and occasionally exhibit a black scrawl. Owing to persecution the Jay is local with us and to the northward only reaches Inverness-shire; the typical form, moreover, is now held to differ from the British, which is again a close ally of other European, Asiatic, and north African species.

Not many years ago the pretty long-tailed black and white Magpie (*Pica pica*) was often seen in most of our counties, and was well known for its cunning ways and jarring notes. Now, however, the balance of nature has been so much disturbed by game-preservers that the bird is becoming rare, except where the preservation is incomplete. No doubt its fate is more or less merited, as it is destructive to young birds and eggs, but it used to be a great feature in the landscape, being continually seen moving with strong but low flight from one shelter to another, or in spring busy round its wonderful nest. This is a great roundish mass of sticks lined with clay and then with roots, while the top or roof is comparatively thin and flat, and access is gained at the side. The eggs, generally more than six in number, are greenish white with olive and brown markings. If we ignore various forms that have been described, the Magpie ranges over Europe, north Asia, north Africa and even western North America, but there are fairly recognisable sub-species.

Magpie

The genus *Corvus*, or Crow, includes not only the Crows proper, but also the Jackdaw and the Raven, all of which are big glossy black birds. The Jackdaw (*C. monedula*) breeds throughout the Palæarctic region as far eastward as the Yenisei river, with the exception of the more Arctic parts of Europe, the Færoes and Iceland; it is easily distinguished by its grey nape, while its note is not a caw, but a repetition of " jake,

jake." The curious tumbling flight, the gregarious habits, and the thievish propensities are matters of common knowledge; the food consists of insects, worms, sheeps' parasites, and so forth, and also of any eggs the bird can procure. Single pairs often choose hollow trees or chimneys for their big stick and wool

Magpie's nest

nest, but colonies are very common and build in holes in cliffs, ruins, church towers, and other like places, as well as in rabbit-holes on the hills. Some half-dozen green or more rarely bluish eggs are laid, with black, brown, and olive blotches or spots.

The Raven (*C. corax*) is now a rare bird in most places, though formerly it commonly bred inland in

92   *Order I*

big trees ; such sites for the nest are practically a matter of ancient history in our southern counties, those now chosen being on hill-side crags or sea-side cliffs, and usually in spots very difficult of access on

Raven's nest

account of overhanging rock-faces. A great mass of sticks lined with softer materials is collected in some larger hole or vertical fissure, and on these are deposited about five green eggs with olive and brown markings distributed over the shell : most exceptionally the coloration is red. They are laid very early in the north, but often later from the Border Country to south-west England. This fine glossy black bird is still fairly abundant in northern Scotland, on the Welsh coast and in Ireland; abroad it breeds throughout Europe, northern Asia and North America, but other species or races take its place in Africa. The flight is powerful but slow, and both sexes tumble in the air, generally in the neighbourhood of the nest; the note is a harsh barking sound, the food is of all descriptions, including weakly lambs, small mammals, birds, eggs, and carrion.

The Carrion Crow (*C. corone*) and the Grey or Hooded Crow (*C. cornix*) may be considered together, for the latter is little more than a grey-backed and grey-breasted race of the former, which interbreeds with it where their ranges overlap. Many migrants arrive in autumn, and do not all leave us in winter. In Britain the black bird reaches from the south to about mid-Scotland, but the particoloured bird is almost confined in the breeding season to the north of the Firth of Forth, the Isle of Man, and Ireland, while it also has a more northerly and south-easterly range in Europe. This, however, must be taken as a very rough statement, as a great amount of overlapping takes place, and both forms extend to northern Asia, while even in Britain nests may be found down to the Scottish borders. Rooks are often called Crows, but the true Crows have a decidedly harsher voice and are, if anything, more difficult to approach; they are notorious egg-stealers and destroy much young game, but they feed also on other small birds and mammals, on carrion, insects, and fish, when they can catch it. They are therefore a public nuisance and not beneficial, as Rooks are. From autumn to spring numbers are seen in the open country or on the shore, but in spring our native stocks retire to the woodlands or the hill-country, where they build on a tree or rock a large nest of sticks, lined with wool and other soft materials. The four or five eggs resemble those of the Raven, but are smaller. On the sea-coast the nest is often in a cliff. Colonies are not formed. The flight is similar to that of the next species.

The Rook (*C. frugilegus*), on the other hand, breeds in colonies often of hundreds of birds, the larger rookeries being also used as winter resorts. The nests are usually on lofty trees, but in quiet places may be much nearer the ground and exceptionally on it; they are built of sticks and mainly lined with straw, while the eggs are "small editions" of those of their congeners. In the rookeries a vast amount of cawing is always going on, and the birds are tame enough, but at other times they are very wary, and constantly post sentinels when feeding in the fields. The amount of insect pests that is consumed must be incalculable, and the Rook is doubtless of the greatest utility, but in certain places it takes to bad habits and imitates the Crow in the destruction of eggs of game. This simply points to the fact that the natural food-supply will only maintain a certain number of individuals, and that any excess should be checked by shooting the young. This species is found breeding throughout Britain as far north as the Orkneys; it does not do so in southern Europe, but ranges over the northern portion and to Siberia. An adult Rook is characterized by a broad tract of white warty skin round the base of the bill, but the young have that part feathered, and are best distinguished from Crows by the dark or livid, as opposed to pale flesh-coloured, inside of the mouth. These

Rook

birds are constant migrants to and from the Continent. The flight is powerful, if slightly laboured.

Family ALAUDIDÆ, or Larks

The Skylark (*Alauda arvensis*) is too well known to need any description, but it must be noted that its erectile crest has sometimes caused it to be taken by the uninitiated for the very different Crested Lark. Whether it be seen rising from the ground and flying low before us, diligently dusting itself on the roads, or soaring high in the air, it is always the same familiar friend, while its joyous song, uttered on the wing at all times of year, is one of the most delightful sounds of the country-side. The notes are no doubt most

Skylark

perfect in the breeding season, when the male rises as far as the keenest sight can follow him and serenades his sitting mate, but they may generally be heard in sunny weather even in winter. The food consists chiefly of insects and worms, with a certain proportion of seeds; the nest is made of grasses, and is built in pastures, young corn fields, or rough herbage, or even on banks and sandy flats, and generally contains four or five eggs of a whitish ground-colour thickly dotted with brown and grey. The Skylark is a great migrant and is constantly seen passing backwards and forwards to the Continent on passage, while it breeds throughout the Palæarctic region (in several forms), and even beyond the Arctic Circle northwards.

The Wood-lark (*Lullula arborea*) is easily distinguished from the Skylark by its smaller size, short tail and much broader buff streak over the eye, or at a distance by its lighter appearance and somewhat Chat-like movements. It does not frequent the interior of woods, but is to be found on dry sunny fields or banks on their outskirts, heaths and sandy places, where it makes its nest in rather bare spots and lays four or five eggs, generally with brighter markings than those of its congener. They are small and vary considerably in colour. The sweet song of reiterated notes is uttered while the bird is hovering at a moderate height in the air, or while sitting on the outer branch of a tree; it is very noticeable in autumn, as well as spring, and would be more generally admired were the songster not so local, for it is never abundant, though found in small colonies in England, chiefly in the eastern and southern districts, in Wales, and Wicklow in Ireland. In Europe and north-west Africa this species breeds in the temperate regions, as it does in Persia and Transcaspia, but it is equally local or but little commoner than with us, while it hardly changes its quarters in Britain during the year. In Scotland the Tree-pipit is often termed "Wood-lark," but the former bird does not arrive till the latter has begun to breed, while the habits, nest and eggs are quite unmistakeable. The food is similar to that of the Skylark.

The Crested Lark (*Galerida cristata*) has very rarely occurred in Britain, though not uncommon across the Channel in Normandy, while it has an extensive range abroad, which cannot be defined in our limited space, owing to the numerous races for which different scientific names have been proposed. The habits, nest and eggs are more similar to those of the Wood-lark than of the Skylark, but the bird itself is more like the Skylark, though it has a much longer crest, rufous colouring under the wing, and no white on the tail.

The Shore-lark (*Otocorys alpestris*), a pretty brown bird with black and yellow head, black chest, white belly, and, in the male, a tuft of black feathers on each side of the head, must now be included in our more or less regular winter visitors, though formerly it was hardly known in Britain. It comes in larger or smaller flocks, chiefly to our eastern shores, but seldom penetrates inland. This bird has been divided into an immense quantity of species or subspecies, from northern Europe, Asia and Africa, North America and even the Andes. Our race breeds in the far north, the warmly lined nest with its brown-spotted eggs being often built in rather stony ground near the shore or on hill-slopes. The song, flight, and food are comparable to those of the Skylark.

ORDER II. PICARIÆ

This order includes the following seven families, which are grouped together on anatomical grounds, as in the case of the *Passeres*. But, whereas the latter are closely linked together by their general appearance and their powers of song, the Picarians, apart from

Order II. Picariæ

their anatomy, are remarkable for their diversity of style and habits. They approach the *Passeres*, however, in being birds of the trees or air rather than of the ground, and in having blind and naked nestlings, except in Goatsuckers and Hoopoes. Swifts have very short beaks with a wide gape, short feathered feet, long and often forked tails, and toes which all point forward. Goatsuckers have an even wider gape, with conspicuous bristles along the upper mandible, a toothed claw on the mid-toe, and peculiarly soft plumage. The Woodpeckers have a big head, with powerful bill and long extensible tongue, stiff spiny tail-feathers, and two of the four toes pointing backwards; but the Wrynecks, classed in the same family, have soft tail-feathers. Our Kingfisher has a large head with a long stout bill, and a very short tail, while it is noted for its brilliant colours. The Roller and Bee-eater, rarely met with in Britain, are equally brilliant, but have long curved bills. Hoopoes have a long slender arched bill and a wonderful compressed crest of feathers. The Cuckoo has its well-known call to distinguish it, and also two toes pointing backwards, as in the Woodpecker tribe, not to mention its habit of laying eggs in other birds' nests. The name *Picariæ* is derived from *Picus*, a Wood-pecker.

Family CYPSELIDÆ, or Swifts

The Swifts are remarkable and interesting birds, which used to be classed with the Swallows, but on investigation proved to be more nearly akin to the Humming-birds. This is a good case of the "convergence" of species, even of different Orders, in nature, that is, of those which stand far apart in classification and presumably in origin, becoming similar to one another for all practical purposes. Apart from accidental visitors, we have only in Britain one member of the family, the black Swift or Deviling (*Micropus apus*), which arrives a little later than the Swallows, charms us by its aerial evolutions, startles us by its piercing screams, and disappears very regularly by September. It is peculiarly a denizen of the air, which

Swift on nest

rises from the ground with difficulty, and, having all its toes pointing forward, can hardly be said to perch. The food consists of insects taken on the wing, as the birds circle widely in the air, singly or in small parties ; the flight is most characteristic, a number of quick vibrations of the wings being followed by periods of inactivity, when the bird glides with motionless wings held in the shape of a bow. Swifts generally breed in small colonies under the eaves of houses, but they also make use of crevices in cliffs and quarries, as well as holes and wall-tops inside church towers and other buildings. The two eggs have the same long outline and roughish white shell as those of Humming-birds. The nest is of a little straw stuck together with saliva. This species is locally common in Britain and in the western Palæarctic region, where several allied species complicate the question of range.

Goatsucker

Family CAPRIMULGIDÆ, or Goatsuckers

The Nightjar or Goatsucker (*Caprimulgus europœus*) is an uncanny looking bird with uncanny ways. The great wide gape with bristles along the upper mandible gives it a curious appearance, though the soft grey, buff, and brown plumage is distinctly pleasing to the eye, while the jarring sounds uttered when it is stationary and the sudden clap of the wings in flight create quite a ghostly impression in the twilight, when it comes out to hawk for the moths and other insects on which it feeds. The female is said to make a less jarring sound; she lacks the white spots on wing and tail. Country-folk consider it a kind of Hawk, and keepers can with difficulty be persuaded how harmless and beneficial the bird is. In the day-time the Nightjar is seldom seen unless flushed from cover, but occasionally it suns itself on some bough or fence, along and not across which it crouches. It comes to us as late as mid-May and leaves about September, during which months it is somewhat local, as it requires wild woodlands, gorse-covers, heather, or bracken in which to breed. No nest is made, but two pretty white eggs mottled with brown and grey are deposited in a depression of the soil. In suitable localities abroad it ranges in various forms from temperate Europe to north Africa and mid-Siberia. The ancients fancied that the bird sucked goats dry, and called it by names equivalent to our term Goatsucker.

Family PICIDÆ, or Woodpecker Tribe
Subfamily Iynginæ, or Wrynecks

The Wryneck (*Iynx torquilla*) is another soft-plumaged species with grey, brown, and buff coloration, which is closely allied to the Woodpeckers. It is often called the Cuckoo's Mate, as it arrives simultaneously in April, but it is not now so well known as formerly, as it is decreasing in numbers in many places. In the north of England it was always rare, while it has never been proved to have bred in Scotland or Ireland. Abroad it is less common in southern than northern Europe in summer, though it ranges to Japan, and a southward migration begins, at least in Britain, in September. The Wryneck, so-called from its habit of twisting its head upon its shoulders, is a quiet bird, which makes short undulating flights from tree to tree, and feeds on the insects to be found on the bark or on ants which it picks up from the ground with its long tongue ; its spring note is very loud, and consists of a single syllable sharply repeated. It does not cut a hole for itself as a Woodpecker does, but lays six or more white eggs in a hole in a tree, normally without any bedding. Sometimes a hole in a bank is selected and more often an artificial nest-box. When disturbed while incubating, the hen hisses loudly.

SUBFAMILY **Picinæ**, OR WOODPECKERS

Three Woodpeckers breed in Britain and all are resident species. Of these the Great Spotted Woodpecker (*Dryobates major*) is found in one form or another throughout the Palæarctic region, except north Africa, and is locally common in England. Large numbers come in from the Continent towards the latter part of the year, and to these is ascribed the present increase of breeding birds in Scotland. An ancient colony which had existed on Spey-side died out, but within the last fifty years a good many pairs have taken up their summer quarters in the Lowlands, have spread across the Border, and have effected a junction with their relations in England. It might be supposed that a black and white bird with crimson on the nape and vent would be very conspicuous, but such is not the case, for the bright colours are not generally evident to the naked eye in woods, while the hen lacks the crimson nape. The food consists chiefly of insects,

Woodpecker's nesting-holes

though nuts, acorns and other fruits are also eaten. When a Woodpecker is searching the bark of a tree it ascends—or even descends—the trunk spirally as a Creeper does, moving in jerky fashion and supporting itself by the stiff-pointed feathers of its outspread tail. The flight is characteristically undulating. The notes are usually sharp and disconnected, but when rapidly uttered produce a sound not unlike the "laugh" of the Green Woodpecker. In spring the bird makes a loud hammering noise by striking the bark, generally of some rotten bough, with its beak. Rotten trees are also chosen, as a rule, for the nesting-hole, which is bored inwards for some few inches in the form of a circle, and is then gradually enlarged downwards until it ends in a spacious chamber, where from five to seven glossy white eggs are deposited on a few chips. Abortive borings are common.

The Lesser Spotted Woodpecker (*D. minor*) is much more local than the last-named in Britain and seldom more abundant; it rarely strays to Scotland or Ireland, but has a similar foreign range, with the addition of north Africa. In this bird the crown, and not the nape, is crimson, and the remaining parts are black and white; but it should be noticed that in both species the young have some red on the crown. Allowing for its smaller size the Lesser Spotted Woodpecker resembles the Greater Spotted in its habits; the eggs are not so large, but the habit of drumming is more noticeable.

The Green Woodpecker or Yaffle (*Picus viridis*), our largest and most plentiful species, is green with a crimson crown and nape, and a black face which has a crimson stripe in the male. It rarely breeds in the north of England and never in Scotland or Ireland, but it extends through Europe to Persia, though south of the Pyrenees it is represented by a grey-cheeked

Green Woodpecker

congener. Its loud laughing cry is said to portend rain, and gains it the name of Rainbird, while in its habits generally it is similar to our other Woodpeckers. It is, however, more frequently seen on the ground, feeding on ants and so forth, and commonly bores its holes in living trees. Heaps of chips may be found below them, which is not usually the case with our other species.

Kingfisher

Family ALCEDINIDÆ, or Kingfishers

Our Kingfisher (*Alcedo ispida*), which is represented outside of Europe by various closely allied species, and in it hardly reaches north of the Baltic, is perhaps the most brilliant of British birds. Not only are its blue, green, and chestnut tints wonderfully beautiful, but it is characterized by a long and enormously strong bill, very short wings and tail. After having decreased during the last century its numbers are now again increasing, and it can hardly be called uncommon in suitable situations, save in north Scotland and Ireland. The Kingfisher may be seen on lakes, rivers and ponds, but is specially partial to our smaller streams or ditches, where it dashes past us like a flash of blue, uttering a shrill reiterated cry, or sits patient but alert on some jutting branch, ready to dart out at a moment's notice on the minnows and other small fish, for which it plunges into the water. It also eats small crustaceans and insects. Six or more round glossy white eggs are laid early in spring in an enlarged chamber at the end of a tunnel bored into the perpendicular bank of a stream or dry pit; the entrance hole is circular and placed out of reach of rats. The egg-chamber is lined with vertebræ and small bones of fishes.

Family UPUPIDÆ, or Hoopoes

The Hoopoe (*Upupa epops*), as well as the Kingfisher, entered largely into the mythology of the ancients. The former would probably breed regularly with us in many places, if not shot in spring; for it often occurs in south England, and has actually nested in most of the counties bordering the Channel or near it. But so curious-looking a fowl, with laterally compressed erectile crest and long curved bill, and with plumage prettily variegated with black, white, and fawn-colour, is a great temptation to the gunner. Abroad it is found in the Palæarctic region except the far north and eastern Asia, while it has near allies in other countries. The Hoopoe is a confiding bird, which flits along with wavering flight or struts about the ground, searching for its food of worms and insects; the note is a soft "hoop-hoop"; the nest of a little grass lined with dung is placed in holes in trees or walls and contains some half-a-dozen pale greenish blue eggs.

Hoopoes

Family CUCULIDÆ, or Cuckoos

Our familiar Cuckoo (*Cuculus canorus*) needs no description, though we may remark that a brownish grey bird closely barred on the lighter under surface is not unnaturally mistaken for a Hawk by the unlearned, especially when it has a similar flight. But the Cuckoo is only with us from April till August, or in the case of the young till a month or two later, and its note is unmistakeable, though that of the hen-bird is merely a bubbling sound. Abroad its summer range extends from about the Arctic Circle to north Africa, Japan and the Himalayas, but it migrates to the extreme

Cuckoo

south of Asia and Africa. In Britain it is almost more common in the north than in the south, for there the insects and caterpillars which form its food are everywhere plentiful, and the Meadow-Pipit is even a more usual foster-parent than the Sedge or Reed Warbler.

Picariæ 111

The eggs are also commonly laid in nests of the Hedge-Sparrow, Pied Wagtail, and Robin, or less frequently elsewhere. With us they are seldom coloured like those of the foster-parent. Two may be found in the same nest, but seem always to be the produce of different hens, if we judge by their colour. It is now certain that the hen Cuckoo often deposits her egg in a nest with her bill, but it does not follow that such is always the case. For the habits of the young Cuckoo, its way of thrusting its nest-mates out to perish, and the curious hollow in the back, so well calculated for the operation, which fills up in less than a fortnight, our readers must turn to the very full accounts in larger books, which will well repay perusal.

ORDER III. STRIGES

This Order contains only one Family, that of the Owls. They have big heads, with thin necks, and large staring eyes; the bill is rather short and stout and is sharply hooked; the feet are short and strong, with a reversible outer toe; the claws are long, curved, sharp, and not uncommonly feathered; the wings are broad and rounded; a cere or fleshy skin surrounds the base of the bill, though it is often covered with feathers or bristles; the upper eye-lid shuts over the eye, and not the lower as in birds generally; the feathering is soft and fluffy, with a curious facial disc round each eye, in most cases quite pronounced; the female is larger than the male. Owls are more or less nocturnal or fly at dusk; they feed on small mammals, birds, and insects, and commonly lay white eggs in holes in trees; the young are helpless and woolly.

Bones, beetles' wings, and so forth are cast up in the form of pellets.

Family STRIGIDÆ, or Owls

We have four native species of this Family in Britain, not including the recently introduced Little Owl, which resemble one another in their soft noiseless flight and in their food, but differ in voice and many of their other habits. The Barn or Screech Owl (*Flammea flammea*), which is sometimes on structural grounds placed in a separate Subfamily, is mottled with brown, grey, white, and bright buff above, and has the face and lower parts white. But there is another phase which has much greyer upper parts and buff under surface, these differences being still more pronounced abroad, where the bird, under one or another subspecific or specific name, extends over nearly the whole world. It is not, however, an Arctic species, while in Britain it is becoming rarer, even in suitable places, owing to persecution. In the gloaming it has a ghostly appearance as it passes with its noiseless flight in search of the small mammals and insects on which it mainly feeds, while its weird screech adds to the uncanny effect. As it only comes out at nightfall it comparatively seldom takes young birds, but is, of course, very beneficial to agriculturists. This "White" Owl lays its five or six dull white eggs in towers, barns, dovecots, hollow trees, holes in cliffs or steep banks without any nest; they are sometimes deposited in pairs at intervals, so that the young may be of very different ages.

The Long-eared Owl (*Asio otus*), on the other hand, is entirely a woodland bird, which breeds in old nests of Magpies, Crows, and so forth, and lays its shining white eggs on a slight bedding of little twigs and feathers, some of which generally drop to the ground and betray the site. When disturbed the parents will sit on a neighbouring tree and snap their bills loudly at an intruder. The cry is shrill and like that of an infant; it is most noticeable after the breeding season, when the young commonly frequent gardens and copses near villages. The colour of this Owl is darkish brown, varied with grey and buff, the lower parts and face being buff, the former streaked with brown. It is found throughout Britain and the Palæarctic region generally, except in the coldest portions, and in Abyssinia, and is represented in the Nearctic by a mere subspecies. It may breed as early as February. This bird has two long tufts of feathers on the head, erroneously called "ears," or hardly more correctly "horns."

The Short-eared Owl (*A. accipitrinus*) is much lighter in colour than its congener, though marked with larger blotches, and the tufts on the head are much shorter. It migrates to us in autumn in large numbers, and is known to shooters from its wavering flight as the "Woodcock Owl." A few pairs, however, which may or may not be resident, breed in Orkney, Cambridgeshire and Norfolk. From our other Owls it differs in being mainly a diurnal bird, which quarters the ground in the day-time for mice, rats, reptiles and large insects; moreover it makes a fairly solid nest on the ground in heather, sedge, or rough pastures. Its foreign range extends not only over the Palæarctic region, but also the Nearctic and Neotropical, excluding the colder parts. In times of vole plagues, as on the Scottish

Border in 1890–1891, vast numbers of these Owls remain to breed, and lay an unusual number of eggs, a dozen

Tawny Owl

having been found in one nest. In the cold season the birds wander to all parts of the globe.

The Tawny or Wood Owl (*Strix aluco*) is our

largest and commonest species. It is varied with brown and grey, with bars on the tail and white marks on the wings, while the lower parts are whitish with brown streaks. It always prefers woods, where towards evening its well-known hoot, " too-whit, too-whoo," resounds through the air; it lays four or more big shining white eggs in a hole in a rotten tree, in the deserted nest of some large bird, or in a burrow, and exceptionally in unused buildings. It is nocturnal, or rather dusk-loving, and remains with us all the year, breeding even earlier in spring than the Long-eared Owl. Curiously its occurrence has not been authenticated in Ireland, nor do the treeless Shetlands or Orkneys suit its habits. It breeds in most of the Palæarctic region, but east of Persia and Palestine we find different species of the genus.

The Little Owl (*Carine noctua*), well known to the Greeks as the emblem of the goddess Athene, is now abundant from Northamptonshire to Cambridgeshire and Essex, and occurs elsewhere. It was first introduced from the Continent in 1843 by Waterton, but his attempt to acclimatize it was a failure. Mr Meade-Waldo made a further trial with greater success, and Lord Lilford in 1888 completely succeeded near Oundle. Formerly the bird was a rare visitor, now it is spreading almost too rapidly, especially where pollard trees are plentiful, in which it usually lays its comparatively small eggs, four or more in number. But it also takes possession of holes in chalk pits and quarries. It is a diurnal bird to some extent, but its noisy cry is chiefly heard in the evening, and in this country its repeated grotesque actions are little in evidence. Its foreign range only extends from the Baltic to the Mediterranean, if we accept as different the Asiatic and African forms. The Little Owl is a small brown bird, spotted with white above and barred on the tail; the face is greyish and the lower parts are white with brown streaks.

Griffon Vulture

ORDER IV. ACCIPITRES

This Order comprises the Vultures, Eagles, Falcons, Hawks, Buzzards, Kites, Harriers, and the Osprey. The Vultures have more or less downy heads, a ruff of feathers over the shoulders, a horny " cere " at the base of the bill, and comparatively straight blunt claws,

Order IV. Accipitres

but only two species, the Griffon and the Egyptian Vulture (p. 259), occur in Britain, and these as irregular and extremely rare visitors. For the rest of the Order the characteristics are as follows :—The bill is strong and decidedly hooked, with a membranous cere, being of the description usually termed " raptorial " or predatory ; the feet, used for grasping prey, are strong with sharp curved claws ; the true leg is often long and conspicuous; while the feet may be feathered to the toes, the fourth being reversible in the Osprey, as in the case of the Owls : over each eye is a bony ridge, which gives the birds a fierce appearance ; the wings are either long (in the forms called by falconers " noble ") or short (in the " ignoble "). The female is larger and finer than the male, though she performs the duties of incubation. The members of this Order have a large crop and are mainly carnivorous, but they also feed on insects; they are land-birds and fliers by day, while the woolly nestlings remain for a long time in the nest, particularly in the case of Eagles. When fledged the young seldom have the transverse streaks common on the older birds, but shew longitudinal markings instead.

Family FALCONIDÆ, or Raptorial Birds

Harriers are now rare birds in Britain, though their migratory habits make it possible that they would re-occupy many former haunts, if properly protected, for they were not uncommon of old. The Marsh Harrier, or Moor-buzzard (*Circus æruginosus*), in particular can now hardly be considered more than an accidental visitor, for it only breeds in a few spots in eastern Ireland, unless a chance pair escapes destruction on the Norfolk Broads. It was always rare in Scotland, but is sufficiently plentiful in the area lying between south Sweden, west Siberia, Kashmir, Egypt and Morocco. Even twenty years ago a pair might generally be seen circling high in the air over the larger waters of the Broadland, regularly quartering the

Marsh Harrier

marshes for prey, or swooping boldly down to their nest among the sedges and reeds. The four or five bluish white eggs are deposited on a large and rather flat mass of reeds or marsh herbage. The food consists of small mammals and birds, frogs and reptiles, with birds' eggs to vary the diet; the cry is shrill, but little heard. In colour this species is of a rich brown, with cream-coloured head and lighter under parts both streaked with brown, the wings and tail being chiefly grey. The female has the tail and under parts brown.

The Hen-harrier (*C. cyaneus*), on the other hand, has the sexes perfectly different. The male is blue-grey with white rump and under parts, while the female is brown with some white on the nape and rump, and streaky lower surface. The tail is very distinctly barred, and thence arose the name of "Ring-tail," for the hen was formerly considered a different species from the "Kite" or cock-bird. Abroad the range differs from that of the Marsh Harrier in that the Hen-harrier does not breed numerously to the south of the great mountain ranges of Europe, though it extends to Kamtschatka and has a very nearly allied representative in North America; in Britain it is now very rare even in its last resorts in Sutherland and the Orkneys, though it used to breed as far south as our eastern fen-lands. This bird makes a less substantial nest than its congener and sometimes has a few rusty markings on the eggs; with us it now breeds only on heathery moorlands, though on the Continent cornfields and marshy spots are commonly preferred. It flies more lightly than the Marsh Harrier and has a more screaming cry; otherwise the habits of the two species are very similar.

Montagu's Harrier (*C. pygargus*), first distinguished by Colonel Montagu, is a much slighter bird than the Hen-harrier, with a peculiarly light and graceful flight. The facial ruff, which all our Harriers shew more or less, is distinct, the lateral tail-feathers are heavily barred, and the white rump of the male Hen-harrier is conspicuous by its absence. Females are harder to distinguish. It migrates to us in spring, and a few pairs nest in the fens, gorse-covers, or heath-lands of England and Wales, chiefly in the southern and eastern counties. Abroad it ranges further south in Europe than its congener and breeds in north Africa, but it is not found so far east in Asia. The nest is generally more flimsy, and the four or five eggs can always be recognised by their smaller size.

The Common Buzzard (*Buteo buteo*) used to be really common in many parts of Britain, and even now is constantly to be met with on the west from Cornwall to Sutherland, though rarely in Ireland. Its foreign distribution is from Scandinavia and the Baltic provinces of Russia to south Europe; eastwards and southwards it is represented by the closely allied African and other Buzzards. Our species exhibits many dark or light phases, but the typical bird is in both sexes brown with some light colour on the breast and a barred tail, while the young are particoloured below. The flight is heavy but strong, and both parents may be seen to advantage when circling round their breeding haunts and uttering their cat-like mew; they are not shy at the nest, which is built in a tree or recess in a cliff, and seems invariably to have a lining of fresh green leaves on the top of the mass of sticks. The three or four thick shelled coarse-grained eggs have a greenish white ground-colour, and either a few brown markings or fine red-brown and lilac blotches; they vary, however, extremely. This is a most useful bird to the farmer or gamekeeper, for it feeds almost entirely on rats, mice, frogs, reptiles, and large insects, and very rarely molests a bird.

The Rough-legged Buzzard (*B. lagopus*) is normally a lighter coloured bird than the Common Buzzard, particularly on the head, tail, and under parts, but the chief distinction lies in the feet, which are feathered down to the toes. In some years regular incursions of this species take place, especially on our eastern coasts, but it also occurs in autumn as an irregular but fairly common migrant from its breeding quarters in Scandinavia and north Russia, whence it extends across Asia and even to Alaska. The habits, allowing for a more northerly habitat, are the same as those of its congener; it is more apt to inhabit treeless country, and has a somewhat stronger flight.

The Golden Eagle (*Aquila chrysaëtus*) is a bird of the mountains, and is found also, in many districts, in forest country. It is distributed over the whole area lying between Lapland and Japan, Spain, north Africa, and the Himalayas, and breeds in North America, though in Britain it is now restricted to the highlands of Scotland, where it is by no means uncommon. Up to the middle of last century the nest might be found in the lowlands, and some century and a half previously in Derbyshire and Wales. The Cheviots and Lakeland Hills were occupied a good deal later. The nest is usually in a big cliff or in some smaller rock up the hill valleys, but abroad, and occasionally in Britain, a tree is chosen to support the huge mass of sticks, which is lined with grass, or in Scotland with Wood-rush (*Luzula sylvatica*); the two eggs are rarely pure white and vary much in the richness of their purplish red, rufous or lilac markings. This Eagle preys upon weakly lambs, fawns, hares, rabbits and other mammals, birds, and even carrion; its flight is very powerful, as may be well seen when it pursues a grouse; its cry is shrill with a harsher termination.

The coloration is rich brown with a fulvous crown and nape and a little grey or white on the tail, while the feet are feathered to the toes.

The White-tailed Eagle (*Haliaëtus albicilla*) still breeds in Shetland, but has gradually disappeared

Golden Eagle

from the north and west of the Scottish mainland and from Ireland; for unfortunately it is fond of carrion, which can be easily poisoned, while it takes young lambs, in addition to any sort of flesh, fish, or fowl obtainable. It nested even in the early years of last century in the Lake district and the Isle of Man, while abroad it does so in Greenland, north Europe and Asia, and from Scandinavia by way of Germany to Turkey and Lower Egypt. It differs from the Golden Eagle in its entirely white tail and bare feet, while the head is streaked with brown and white. Its flight is even stronger and its cry more of a yelp. The nest, sometimes built on the ground in marshes or on islands in lakes, with us now is in precipitous cliffs, and may be composed of sea-weed where sticks are scarce. The two eggs are white. By far the greatest number of the birds that stray to England are White-tailed and not Golden Eagles.

The Sparrow-hawk (*Accipiter nisus*) is a dashing marauder whose short wings and long tail combine to produce easy steering with rapid flight; it preys upon birds caught as it dashes along the fields and hedgerows, and eats them upon the ground, where it leaves little heaps of feathers. Its cry is sharp, its nest a large flat structure of sticks and twigs having a depression in the centre to contain the four or five greenish white eggs boldly marked with reddish or blackish brown, which have a peculiarly thick shell. The site chosen is most commonly in a fir-tree. This well-known bird is found throughout our woodlands, as well as over the whole Palæarctic region. It is grey-blue above and buff below, the tail and under parts being barred, and the cheeks rufous; the female has usually a greyer breast and rufous flanks.

The Goshawk (*Astur palumbarius*), formerly much used in falconry, used to breed in northern Scotland until the end of the eighteenth century, while in England it was once the custom to turn out the old birds to nest in the woods, with a view to capturing the young subsequently. It is now a very rare visitor in the colder months, though it is by no means uncommon in most parts of the Palæarctic region. In Scotland the Peregrine Falcon is often called Goshawk, a fact which may give rise to erroneous records. Being an ashy brown bird with barred tail and white under parts thickly barred with black, this species has a general resemblance to a large Sparrow-hawk, and is similar in its habits, but the eggs are bluish white and rarely have even faint rusty markings. The food consists of mammals and birds.

The Kite (*Milvus milvus*) is one of our most interesting survivals. Long ago it was the common scavenger of our towns, and probably everyone has heard stories of its boldness in snatching food from children's hands, and so forth. Now a few pairs maintain a precarious existence in and near Wales, though it is hardly more than a decade since others bred in Scotland, while so few individuals visit us from abroad that we can hardly hope for an increase from that source. A large rufous bird, with streaks below and on the white head, must always be conspicuous, while its long deeply forked tail, well seen as it makes bold circles in the air, and its shrill mewing cry, render it an object that cannot easily fail of observation. Besides offal the diet comprises small mammals, birds, frogs, and even fish; the nest of sticks, paper, rags and other materials is placed in the fork of a tree, or more rarely in a cliff, and contains three eggs, similar to, or a little duller than, those of the Buzzard. It ranges from Europe and north Africa to Palestine and Asia Minor, but is less common in Scandinavia and north Russia.

It is impossible to say, until each summer comes

round, whether the Honey Buzzard (*Pernis apivorus*) still breeds with us. It is an immigrant from abroad, not arriving at the earliest before the end of May, and is difficult of observation in leafy June. Undoubtedly it used to visit us regularly, while a few pairs bred in the New Forest and, by chance, as far north as Ross-shire; but the greed of collectors has almost exterminated the bird, and in any one year it is impossible to predict for that following. It is a local species distributed sporadically in summer from about the Arctic Circle to the Mediterranean, across Asia north of the Himalayas and in a slightly different form to Japan; its flight is fairly strong and its actions on the ground easy, as might be expected from its habit of feeding chiefly on insects, specially wasps, bees and their larvæ, in addition to slugs, worms, small mammals and birds. The shrill cry of the adult is little heard. The nest of sticks is lined with fresh leaves; the two or three eggs are exceptionally round, yellowish white, and finely blotched or almost completely covered with bright red or reddish brown. The colour of the Honey Buzzard is brown, with ash-coloured head and whitish lower surface, the latter being streaked and the tail barred with brown.

Our next three species are particular favourites of the falconer, but are merely occasional migrants to Britain, and, as they come from the far north, naturally occur for the most part in Scotland. The true Greenland Falcon (*Hierofalco candicans*), a circumpolar species which breeds north of the Arctic Circle, is invariably white with black streaks and spots above; the Iceland Falcon (*H. islandus*), confined to Iceland, south Greenland, Jan Mayen and north Siberia, is brownish grey and cream-coloured above, and marked with brown below; the young, however, are browner and often very difficult to distinguish from those of the Scandinavian Gyr Falcon (*H. gyrfalco*). This is a somewhat smaller bird of a more lilac grey colour above, which should have the head much darker, the streaks below more numerous, and a black cheek-patch, as in the next species. These three large northern forms are similar to the Peregrine

Greenland Falcon

Falcon in flight, food, and cry, while they lay their eggs in old nests—probably of Ravens or Crows—or on ledges of cliffs without any nest. The eggs, however, are bigger and lighter in the markings.

The Peregrine Falcon (*F. peregrinus*), the falconer's mainstay and the "Hunting Hawk" of country-folk, is the fiercest of our resident members of the Family, for our eagles are comparatively timid and the stories of their ferocity should be credited to other birds. If we were to ignore closely allied species and subspecies, such as the Barbary Falcon of the Mediterranean and the Duck-hawk of North America, the Peregrine might be said to occur throughout the world at one season

Peregrine Falcon

or another, but our form is Palæarctic and barely ranges to the extreme north-east of America. It is more rufous below than the Gyr Falcon and has more decidedly black cheek-patches, added to which it is smaller. In Britain it breeds locally on the cliffs of the south coast and more commonly both on sea-cliffs and inland rocks in Wales and the north of England, while in Scotland and Ireland it is not uncommon. The Peregrine has a fine dashing flight, and the male comes so close to intruders upon his nursery as to be really dangerous, for the rich red-brown eggs are deposited on some narrow ledge or in some little angle of the rock, where the foothold is very precarious. The cry is of a shrieking nature, and the food of rabbits, land-birds and sea-fowl. This species cannot manage such large hares as the big northern Falcons, but is the scourge of grouse moors, and therefore exterminated in many places. Occasionally it lays in deserted nests of large birds in trees or on buildings.

The Hobby (*F. subbuteo*) is a dark grey bird with white cheeks and throat, the remainder of the head black, the lower parts buff streaked with black, the thighs and vent rufous. It ranges throughout Europe, northern Asia and north-west Africa, while a few pairs visit different districts of England from mid-May until September. It has been known to nest as far north as Perthshire, while stray individuals may occur at any time. As it breeds in June, this long-winged species is doubtless often overlooked, for the deserted nests of other birds which it utilizes are then hidden by the foliage; the yellowish white eggs are generally closely stippled with orange-red, but are sometimes finely blotched with darker red, when they resemble those of the Kestrel. It feeds largely on insects, but also on small birds, which it pursues with remarkable swiftness; near the nest the parents circle round high in the air, and are then said to look almost like Swifts. The shrill cry is seldom heard.

The Merlin (*F. æsalon*), locally termed the Stone-falcon, from its habit of perching on boulders, breeds from the Shetlands to the Derbyshire moors, in Wales and in Ireland; abroad from the Arctic Circle to the

Merlin's nest and eggs

Pyrenees and throughout north Asia. The nest, a very slight structure, if there be any, is usually on the ground among heather, but not uncommonly a deserted bird's habitation in a tree or on a ledge of a cliff is made use of, the four or even half-a-dozen eggs being rich brownish red, and rarely shewing the ground colour. After the breeding season this species commonly resorts for a time to the coast-lands, while it may be seen in Britain at all seasons of the year. Its flight is rapid with quick turns; its cry is rather querulous and consists of one reiterated note; its food is of our smaller birds, for the capture of which it was used in falconry. The male Merlin is grey with a rufous hind-neck, a dark band on the tail, and streaks on the buff lower surface, the upper parts exhibiting decided shaft-streaks on the feathers. The female is quite different, being brown with five bars on the tail, the tip of which, as well as the hind-neck and cheeks, are white.

The Kestrel (*F. tinnunculus*) is undoubtedly our commonest Hawk, being found from the north to the south of Britain in the breeding season, but moving gradually southwards for the winter, when it is joined by many migrants from abroad. A like distribution holds good for Europe and Asia, but the north African race is smaller and darker. This most useful species feeds almost entirely on small mammals and insects, especially beetles, though it does occasionally take young birds; the flight is slow and steady with prolonged intervals of hovering, from which habit it is called by country-folk the Windhover. The cry is shrill, and usually heard near the breeding quarters. The Kestrel builds no nest, but re-lines with the slightest of materials that of a Crow, Magpie or other large bird; not uncommonly it lays its eggs on ledges of cliffs or buildings, in holes in chalk pits, or in hollow trees, but rarely on level ground. The eggs are reddish white, partly or entirely covered with red-brown blotches, spots, and smears. The male has a bluish head, tail, and rump, a rufous back with black spots,

Accipitres

and lighter under parts with dark streaks; the female is rufous above with black bars, and has also several tail-bars. Pellets of food are often ejected, as in owls.

Kestrel

The Osprey (*Pandion haliaëtus*) has a very wide range in the world, extending over Europe, Asia, northern Africa, and North America, and even reaching to Australia. It still visits us from autumn to spring, but it is to be feared that not a single pair now breeds in Scotland, where eyries were formerly not uncommon. The large nest of sticks and turf, lined with softer materials, is built in a tree or on a rocky islet in a lake, while an old ruin makes a good substitute for a rock, and in America the birds often form colonies. The eggs, two or three in number, are creamy white with fine blotches of purplish or reddish brown and greyish lilac. The Fish-hawk, as it is sometimes termed, lives entirely on fish, and it is a magnificent sight to see a pair of these unsuspicious birds careering over the waters of a lake, uttering their loud screaming cries, and plunging into the water after their prey. They are brown above and white below, the head and upper breast exhibiting both colours.

ORDER V. STEGANOPODES

The members of this Order stand alone among birds in having all the toes connected by full webs, and the first toe turned somewhat forward, while they have stout curved claws. The bill is long and pointed in Gannets, long and hooked at the tip in Cormorants, and is furnished with a big pouch underneath in Pelicans; Cormorants, moreover, have a very long neck. The feet are set far back, so that walking is difficult; the wings are long and ample; and the tail consists of strong stiff feathers. They are all water birds and for the most part marine, though Pelicans and sometimes Cormorants build inland. They breed in colonies. The blackish nestlings are blind and naked, but soon become covered with white down. Pelicans are not now found in Britain, but their bones have been dug up in Norfolk and Cambridgeshire.

Order V. Steganopodes

Family PELECANIDÆ, or Cormorants and Gannets

The Cormorant (*Phalacrocorax carbo*) is an interesting though ugly bird of a general black hue, but the throat is white, as is a patch on the thighs in spring. It has a very long hooked beak and yellow pouch below it, while its great length makes it a conspicuous object as it passes with low and laboured flight over the sea from one resting place to another. The back

Cormorants and nests

of the head bears a sort of crest in spring, but this is very different from the frontal crest of the Shag, which moreover lacks the bronzy tint adorning the Cormorant's back. Both species have a croaking note and dive deep under water to catch the fish on which they feed. The nest is an immense mass of sea-weed when it is built on islands or ledges of cliffs, but consists of sticks and softer substances when on a tree or bush. Colonies are invariably formed, but are rarely found inland in

Britain, where the bird occurs at all seasons of year on the sea in the most suitable localities. The elongated eggs are light greenish blue with a white chalky covering, and are about four in number. The naked and blind black nestlings feed themselves by thrusting their heads and necks into the parents' bills. Abroad the Cormorant inhabits Europe, Asia, and the Atlantic coast of North America, except the extreme north, and also northern Africa, Australia, and New Zealand, the nesting sites being not uncommonly in marshes and swamps.

The Shag (*Phalacrocorax graculus*) is entirely black with metallic reflexions, and has a frontal crest in spring. Its nest is often on a ledge in a cave, but equally often on a cliff or under a large boulder; the eggs are like those of the Cormorant, but smaller, and are laid rather earlier in April. The flight is hardly so heavy, but otherwise the habits are identical. Locally this "Green Cormorant" is the more common of the two species, especially in the west of Britain, but abroad it is only found from the northern coasts of Lapland by way of Iceland, Norway, and France to Portugal, if we consider the widely spread Mediterranean form to be distinct. By our fishermen the bird is called the Scart or Scarf, a name less commonly used for its congener.

The Gannet or Solan goose (*Sula bassana*), a very local species in Britain, breeds on Grassholm in Pembrokeshire, Ailsa Craig at the mouth of the Clyde, St Kilda, several rocky stacks in the north-west of Scotland, two in the south of Ireland, and the well-known Bass Rock in the Firth of Forth. In some of the northern localities, however, the number of individuals is enormous. The Shetlands, Færoes, Iceland, and some islands in the Gulf of St Lawrence complete the range, but there are other species of Gannet which have a much more southerly distribution. The nest is more grassy than that of a Cormorant, and contains but one egg, from which the bird will hardly allow itself to be dislodged;

Gannets nesting

it is exactly like that of the Cormorant, but a good deal larger. The sitting Gannet uses its long, strong, pointed beak with great effect upon any intruder who disturbs its peace of mind, and utters low guttural or croaking sounds in its distress. It has a strong and steady flight, and when fish are plentiful keeps in companies, while it is one of the finest sights in the world to see the bird dive straight down from a height with closed wings and plunge deep into the water, under which it remains for several seconds. On the Bass Rock, where it was formerly called the Bass Goose, hundreds of birds used to be killed, salted down, and eaten as delicacies. The fully fledged young are sooty black with white markings, instead of being white with black primaries.

ORDER VI. HERODIONES

The Herons and Bitterns are birds of swamps and river-sides, the former often breeding in colonies, called heronries, which are often of considerable size, and the latter leading a skulking life in reedy or sedgy marshes. Herons' nests are usually on trees or bushes, but may be on the ground. The neck, legs and feet are long; the bill is elongated, strong and pointed; the long toes are slightly webbed, with rather short claws, except in Bitterns; the wings are large; crests and ornamental plumes are not uncommon. The nestlings are covered with scanty hair.

Storks have a still stronger bill, that of Ibises is much weaker and extremely curved, that of Spoonbills is flattened into a sort of "spoon" at the end, which finally turns downward. The latter breed in colonies, while all have downy young. One Ibis is an irregular visitor to Britain.

Family ARDEIDÆ, or Herons and Bitterns

The common Heron or Hern (*Ardea cinerea*) is "all length." Its body is thin, its legs, bill and neck are very long, and it has a bifid black crest hanging down from the nape. Otherwise it is grey, with whitish face and lower parts and streaked neck. The

Common Heron

distribution is rather peculiar, for in Europe it only breeds from Scandinavia and Russia in the north to France, central Italy, and the Danube; but in Asia it is found as far south as Ceylon, and in Africa to the Cape. It nests on lofty trees, cliffs, bushes and reeds, or even on the ground, colonies being much more usual than isolated pairs. These heronries are decidedly local in England, but are mostly larger than those in Scotland and Ireland; in Scotland they are both on the mainland and certain islands, while all round the Borders the country is found particularly suitable by the bird. A fine sight is this solitary fisherman as it stalks solemnly through the water, stands for long periods patiently watching for its prey, or makes quick downward darts on the fish with its long sharp bill. Reptiles, frogs, and small mammals form part of the diet, nor are mollusks, worms, crustaceans and insects rejected. The voice is harsh and croaking, with a shorter and sharper alarm-note; the flight is slow but easy, with many a flap of the huge wings, and is often high in the air; the nest is a great flat mass of sticks lined with grass and contains four blue eggs, laid between February and April, but generally early. The Heron was of old a well-known quarry for Falcons, while the young were considered a delicacy for the table.

The Purple Heron (*A. purpurea*) is an irregular migrant to Britain, but breeds not uncommonly in Holland, whence it ranges over west and central Europe and Asia to south Africa. The breast is maroon, the neck rufous with a black stripe on each side. This species is comparatively seldom observed on the wing and has a hoarser note than its congener·

its food, however, is similar, though chiefly sought in the twilight, and its eggs are merely smaller and rather darker in colour. The nest of herbage is built on a flattened mass of vegetation in reed-beds or swamps.

Two other Herons are not very uncommon migrants to Britain, the Squacco Heron (*Ardeola ralloides*) of central and southern Europe, Africa and western Asia; and the Night Heron (*Nycticorax nycticorax*), of a somewhat similar range in Europe, which extends thence over Africa, most of the temperate and hotter parts of Asia, and is represented in North and South America by very closely allied forms. The Squacco Heron is a buff bird with white breast, which has long plumes on the head and back during the breeding season; it is a comparatively small species, and its habits therefore differ to some extent from those of the Common Heron. The Night Heron is grey, with black crown and back, and white head-plumes and breast. The note, generally heard after dusk, is mournful. Both species breed in colonies, but the former prefers low trees and bushes for its nest, while the latter occasionally selects reed-beds.

The Little Bittern (*Ixobrychus minutus*) holds an intermediate position between the Bitterns proper and the Herons, with an inclination to the habits of the former. It used to breed occasionally in the eastern districts of England, but no East Anglian nests have been recorded for many years, and the bird is now only an irregular visitor to any part of Britain. The male is greenish black with buff neck and lower surface; the female is nearly brown above and much streaked with white below. In Europe it nests south of the Baltic, whence it ranges to north Africa, Central Asia, and northern India. The food, of the same description as that of Herons, is generally sought at night, for in the day the Little Bittern lurks in the marshes and, if disturbed, stands silently with upturned beak, looking like a dry reed-stem. Naturally the flight is not very strong, while the cry is of a grunting nature, sharper in the female. The nest of marsh vegetation is placed in reeds, low bushes or even trees and hedges, in or near swamps, the four or five eggs being white with scarcely a tinge of blue.

The booming note of the Bittern (*Botaurus stellaris*) has again been heard in England. Since the days when it bred not uncommonly in the marshes of that country, Wales, and the south of Scotland it has always continued to visit us in spring, but was shot or driven away by persecution in almost every case of later date than 1868. Nevertheless a very young bird was killed in Norfolk in 1886, and finally in 1912 one that had lately left the nest was found on the Broads of that county, while a thorough search terminated by the discovery of the nest itself. The Bittern bred in the same district again in 1913, and there is no reason that it should not continue to do so, if properly protected. Abroad it occupies the whole Palæarctic region, except the far north and north Africa. It is a most striking bird, having very soft plumage, mottled with buff, chestnut and black, while the head is black, and the neck exhibits a fine ruff of erectile feathers. It has a slow flight when seen in the day-time, and is seldom alert except at night, when it hunts for its food, which is as varied as that of the Heron; at other times it skulks in the dense cover of marshes. The nest of dry reeds or flags contains four olive-coloured eggs, often laid in April or even in March.

Family CICONIIDÆ, or Storks

The White Stork (*Ciconia ciconia*) is a rare straggler to Britain, but is too familiar a bird to be omitted here. It is common, and for the most part protected, from Sweden, Holland, and central Europe to Turkey

White Storks

and Greece, as well as in Spain; in north Africa it is abundant; in Asia it ranges eastwards to the central portions and thereafter meets the Japanese species, while it has bred in Ceylon. It is a really white Stork, only the wing-feathers being black, and the bill and legs red. Instead of using their voice Storks clatter their bills loudly; their flight is very strong, and on migration the flocks pass at immense altitudes. Anatomically they are closely connected with Herons and their food is identical, at least in the case of the present species. It breeds on trees, towers, houses or their chimneys, and often on cart-wheels purposely erected on poles; the three to five eggs are white with a coarsely grained shell. The Black Stork is an irregular visitor to us, and will be found in the subsequent list, while the Glossy Ibis comes in the same category, at least nowadays.

Family PLATALEIDÆ, or Spoonbills

Far different is the case with the Spoonbill (*Platalea leucorodia*), which still strays frequently to our shores, notably those of Breydon Water near Yarmouth in Norfolk. This fine white bird, with its black and yellow bill expanding into a flat " spoon " at the end and its red eyes, used to nest until the twelfth century or later both in Norfolk and Suffolk, and four centuries afterwards in Middlesex, Sussex, and Pembrokeshire. There it bred in high trees, often among the Herons, but in Holland, southern Spain, and the Danube region it usually selects reed-beds or bushy trees. It does not now pass the summer in France, but is found at that season in north Africa and across Asia to China and Ceylon. Some half-a-dozen eggs are deposited on a mass of reeds or the like, according to situation; they have a rough white shell, but are prettily marked with reddish brown. The Spoonbill hardly utters any cries, but is not a very shy bird; its diet includes fish, frogs, mollusks, crustaceans, and insects, most of its food being obtained by moving round and round with its bill immersed in the water. The flight is sufficiently strong, without any peculiar characteristics.

Spoonbill and Ibises

ORDER VIII. ANSERES

The Geese, Swans and Ducks are all classified in one Family, *Anatidæ*, and shew a great general similarity, though Swans differ in their much elongated necks. The bill is broad and generally flattened, though more conical in Geese, and comparatively long and thin with hooked tip in the Merganser Ducks; it is covered with soft sensitive skin and ends in a horny piece called the "nail." Both mandibles are furnished at their edges with small hard cross-plates, which act as a sieve when the bill is shut. The feet are short, the front toes webbed, with the hind-toe at a higher level, and the claws small and curved. The wings are rather long and broad; the tail is usually of moderate length, but is much produced in such forms as the Pintail and Long-tailed Duck. For the simultaneous loss of the flight-feathers and the eclipse plumage the reader must be referred to the Introduction. The young are covered with thick yellow, white, grey or brownish down and run as well as swim from the shell. The eggs are unspotted white, green, or creamy, and are bedded in down torn from the female's breast.

Family ANATIDÆ, or Geese, Swans, and Ducks

Not only is this Family very extensive, but the members are distributed over the whole of the globe, while many of them are residents in or migrants to our country. For instance, the King Eider is almost exclusively an Arctic species, the Whooper Swan mainly so, the Steamer Duck is Patagonian and Chilian, the Musk Duck is Australian, and one of the Teals comes from South Georgia in the Antarctic Seas. Though all are in the main anatomically alike, yet they are sufficiently different in appearance and habits to admit of their being placed in eleven groups or Subfamilies, of which five concern us as including British species.

SUBFAMILY Anserinæ, OR GEESE

We have only a single species of Goose which breeds with us, one of our four "Grey Geese" as opposed to our "Black Geese." This is the Grey-lag Goose (*Anser*

anser), which till the end of the eighteenth century bred in the fens of the counties of Cambridge and Lincoln, and somewhat later in the last-named. Now it only does so in Caithness, Sutherlandshire, and Ross-shire—including the Hebrides. Abroad it ranges from Iceland to Kamtschatka and the Danube, while a few

Mute Swan

pairs nest in Denmark, Holland, north Germany, and even as far south as the Mediterranean. In Asia its distribution is less certain, but it appears to reach China and India. The flocks fly high in the shape of the letter V, and, except when actually breeding, several individuals are generally seen together. The cry is often syllabled as " honk-honk," but the bird can hiss like a tame goose. It feeds by day, on grass and other green herbage, as well as on grain among the stubbles; from late summer onwards it resorts to the sea in the evening. The nest of coarse materials, gradually lined with down by the sitting female, contains about five rough-shelled yellowish white eggs, and is built among either grass or heather when the locality permits; at times it is on almost bare soil. This Goose is grey-brown with white belly, the legs and bill being flesh-coloured with a white "nail" at the tip of the latter.

Our other three Grey Geese differ not at all in habits, but have a different distribution. The White-fronted Goose (*A. albifrons*), which has also a white nail on the bill, but a conspicuous white band on the forehead, black bars on the breast, and orange-yellow bill and feet, breeds within the Arctic Circle in most of Europe and Asia, in Greenland and in Iceland; it is accompanied eastwards from Scandinavia by the smaller and darker species or race called the Lesser White-fronted Goose, and in Arctic America by the larger Gambel's Goose. On migration it is more common on our west coasts than our east, while the Scandinavian form rarely visits us. The Bean Goose (*A. fabalis*) and the Pink-footed Goose (*A. brachyrhynchus*) have a black nail on the bill, which has also a black base: in the former bird the central portion of it and the feet are orange, in the latter pink. The colours, however, vary a little with age. Neither species has a white forehead or a barred breast. The Pink-footed Goose is much the commoner on our eastern coasts, breeds in Iceland, Spitsbergen, and probably Franz Josef Land; the Bean Goose from northern Scandinavia, or perhaps only Russia, through the Arctic seas to western Siberia; the latter is locally more abundant in western England as well as in Ireland. To sum up, the "black-nailed" Geese are those which compose most of our large winter flocks of Grey Geese, and among them the Pink-footed in most cases outnumbers the Bean Goose.

Finally we come to our two black Geese, the Bernacle Goose (*Branta leucopsis*) which breeds from about Greenland to Spitsbergen in the Arctic seas, and the Brent Goose (*B. bernicla*), certainly the commonest of the whole group in Britain in the cold season, which in summer ranges from the east of Arctic America eastward to the Taimyr Peninsula in Siberia. A few pairs of the first-named nest on the Lofoten Islands off the coast of Norway. Allowing for an even more northerly range, the habits of the Black Geese are similar to those of the Grey Geese, but they feed on the salt oozes or mud flats rather than on the land, and are very partial to the grass-wrack that often covers the flats. The eggs are more creamy in coloration and smaller, as is natural, for the birds are not as large as Grey Geese. The Brent Goose is entirely black, except for the white belly and a spot on each side of the neck, but the Bernacle Goose differs in having the entire face white and the upper parts lightish grey with white and black bars.

SUBFAMILY **Cygninæ**, OR SWANS

The Swans are known to us all from our domesticated birds. They fly very powerfully, if a little heavily, but spend most of their time swimming about or sleeping on the water, when they are not occupied in feeding. To reach the water-plants which form their main diet they are sometimes obliged to turn tail upwards as ducks do, but their long necks generally obviate this difficulty, and they are not accustomed to dive. The diet may be varied by insects and plants which are not aquatic.

Our three species of Swans are all perfectly white, with black feet, though they differ in size, and are most easily characterized by the bill. The Whooper (*Cygnus cygnus*), which used to breed in the Orkneys and Greenland, but now only ranges from Iceland to Kamtschatka, chiefly north of the Arctic Circle, has a yellow beak with black tip, the yellow colour extending forward to include the nostrils. The much smaller Bewick's Swan (*C. bewicki*), a more Arctic and eastern species, which nests north of the Arctic Circle from north Russia to the Lena delta in mid-Asia, has less yellow on the bill, for it does not reach the nostrils. The Mute Swan (*C. olor*), distinguished by a knob on the forehead, is not a native of England, though many pairs breed on our waters and not uncommonly act as if perfectly wild later in the year. It ranges from Denmark and south Sweden through north Germany to Greece, Turkestan and even Mongolia, and resembles its congeners in its southward winter migration. All three species build huge nests of green plants, grass, and so forth, which are sometimes hidden, but more often quite open to view; the half-dozen or more eggs are very large and white in colour with a yellowish, or in the Mute Swan a greenish tinge. The voices differ considerably, but are, in the wild state, loud and trumpet-like; tame birds are naturally less vociferous.

Unlike Geese and Swans, Ducks vary extremely

both in coloration and habits. It is true that they have many points in common, which we need not repeat in each group, but these groups shew almost as many differences between them as separate the Ducks from the Geese and Swans. In the first place the flight of all the species, when fully on the wing, is much the same, though it may be stronger or weaker; the birds keep more or less in the shape of the letter V, or when coming to rest at night circle round and finally drop down on the water with sudden decision; in the second place the voice is not always a mere quack, as in the tame Duck, and the food, nest and eggs shew considerable variation. The males of nearly all the Ducks have an "eclipse plumage," that is, they become like the females in summer, while the birds lose all their quills at once and become temporarily flightless.

SUBFAMILY **Anatinæ**, OR FRESH-WATER DUCKS

In the opinion of many our handsomest duck is the big Sheldrake or Bargander (*Tadorna tadorna*), which only leaves the coast when driven by storms, except so far as it moves up our tidal rivers towards winter, or occasionally breeds inland near some large lake, such as Loch Leven in Fifeshire. Locally it is found round the whole of the kingdom, haunting sandy shores and links, where it nests in rabbit-holes and lays nearly a dozen beautiful creamy white eggs on a bed of white down pulled from the duck's breast. Exceptionally the bird breeds in drain-pipes, in holes in masonry, or in gorse-coverts. The food consists mainly of mollusks, crustaceans, and insects; the note of the male is shrill and of a barking rather than of a quacking nature, that of the female less piercing. This species ranges locally from Scandinavia and Holland to Japan, and also to the coasts of France and Spain. It has the bill and its conspicuous basal knob red; the head glossy green; the collar white followed by a fine band of chestnut; the speculum green; the remaining parts black and white.

Sheldrake

The Wild Duck, or Mallard (*Anas boschas*), is too well known to need a full description, whether it be of the buff and brown female or of the finely coloured male with his bottle-green head, chestnut breast, and curly tail-coverts. We must, however, notice the "speculum" or bar on the wing, which often serves to determine the species in this Family. In the drake it is purple, edged with white, in the duck it is green.

This species occurs throughout the United Kingdom, and in Europe, Asia, north Africa, and North America, avoiding, however, the Arctic and tropical portions. As it is the origin of a great many of our tame ducks, we are quite familiar with its quack, but it generally sleeps in the day and feeds at other times. It is not particular whether its food is animal or vegetable, seeds and herbage, worms, insects, slugs, frogs, or even fishes being eaten. The nest, built from March onwards, is composed of grass and lined with down; it contains nearly or quite a dozen grey-green eggs, and is placed among rough herbage on the ground; exceptionally a hollow tree or deserted bird's nest is chosen. When a duck leaves her eggs she covers them with the down, if time is given her. In winter our native birds are joined by large numbers of migrants from abroad.

The Gadwall (*A. strepera*), a brown bird mottled in the male with grey, with a chestnut bar on the wing followed by one of white, and with white lower parts, now breeds not only in Norfolk, where it was introduced some sixty years ago, but also in other parts of the kingdom, including even northern Scotland. Except for these breeding individuals the bird is only locally common, though it may occur in any county. It ranges over temperate Europe, Asia, and North America, avoiding as a rule the coldest districts as well as the hottest, but remaining for the summer in Iceland and the south of Spain. This species is almost restricted to fresh water and breeds near lakes and swamps, though usually in long grass, bracken, nettles, rushes or other cover; the eggs are smaller and more elongated than those of the Wild Duck, and of a rich cream-colour. The food is almost entirely vegetable, consisting of leaves, seeds of aquatic plants, and so forth; the note is harsh and rattling.

The male Shoveler (*Spatula clypeata*) is a pretty bird of a general brown and white colour, with a green head and neck, blue upper part of the wing, and chestnut breast, the wing-bar being green. The bluish bill is expanded like a spoon at the tip. The female is almost brown. As a breeding species the Shoveler has multiplied greatly of late years in Britain, and is now found locally in summer even in the north of Scotland. Its numbers increase in the cold season, while in summer it ranges over the northern hemisphere both in the Old and the New World, and reaches northern Africa; it is, in fact, one of the most widely distributed members of the Family. It is a comparatively silent bird, in habits resembling our larger Wild Duck, but it is apt to nest further from the water, while the eggs are smaller and greener.

The Pintail (*Dafila acuta*) is well named the Sea-Pheasant, for its long slender neck and still more its elongated central tail-feathers give it a superficial resemblance to our familiar game-bird. In colour it is mottled with grey on the back and flanks, and prettily marked with fawn-colour, black, and white on the wing, which has a metallic green bar edged with buff. The under parts are white, and there is a white stripe on each side of the bronzy neck. This dainty little species, of which the female is varied with lighter and darker brown above and is whitish below, has of late begun to breed in a few parts of Scotland, including the Shetland Islands; formerly it was a regular but not very common visitor to the coast.

Its foreign range is wide and extends over northern Europe, Asia, and America, but it is difficult to lay down any precise limits, as a few pairs seem to nest regularly in south Europe. The food consists of herbage, insects, and mollusks, but, as the Pintail resorts to mud flats, estuaries, or even the sea in winter, its diet varies somewhat according to circumstances. It is a quiet bird which quacks little, and is said on good authority to have a whistling note as well. Eggs are not found much before the end of May ; they are long in shape and greyish green in colour. The nest is substantial and is generally built near the water.

The Teal (*Querquedula crecca*) is our smallest duck and is plentiful as a migrant in winter, while in summer nests may be found from Shetland to the south of England in suitable places. They are built among heather or rough grass, nearly always at some distance from the water, and therefore are not likely to be found except on moors or uncultivated country. Little ponds, such as suffice for the Mallard, are of no use to the Teal, though its food is similar. The small eggs are cream-coloured with a greenish tinge. The foreign range extends over all Europe and temperate Asia, short of the Arctic Circle, but the bird is chiefly a northern species. The drake's beautiful colours must be seen to be appreciated; here it must suffice to say that the back is prettily marked with black and white and the breast spotted with black on a buff ground, while the head is chestnut varied with buff and metallic green, and the wing-bar purplish and green. This wing-bar serves to distinguish the female from other brown ducks.

The Garganey, or Summer Teal (*Q. querquedula*), only

visits us in spring and autumn, with the exception of a few pairs which remain to breed. It is impossible at present to say whether such instances are more numerous than formerly or not, for they are certainly less frequent in Yorkshire, Northumberland, and on the Norfolk Broads, but the bird has been proved to

Nest and eggs of Teal

nest in the Romney Marshes in Kent, in Somerset and Hants. Abroad it is much more common, ranging from Denmark, Sweden, and Russia to the Mediterranean and throughout northern Asia, while in winter great numbers move down to north Africa, India, and elsewhere. The colour is mainly brown, but a white line descends from each eye to the neck, the wing-coverts are bluish, and the wing-bar green edged with white. Even the brown is varied in the male by delicate lines of white, and the belly is whiter still. The note is rattling and is thought by country-folk to resemble the sound made by a cricket; the food is much as usual in ducks, with a rather greater proportion of fishes; the nest is sometimes placed a little nearer the water than that of the Common Teal, while the eggs are of a rich cream-colour, with no green tint, and resemble those of the Gadwall, though smaller.

The Wigeon (*Mareca penelope*) is a bird of delicate rather than brilliant coloration; the male is pencilled with grey and white on the back and flanks, having a chestnut head with light buff crown, white shoulders—as in the brown female—and a green wing-patch. It is probably the favourite duck of sportsmen, for it arrives in huge numbers towards the end of August and, especially on our mud flats, forms a large part of the bag of a punt-gunner. Though keeping to the sea at this time of year, it breeds over a considerable area in north Scotland as well as on the Border mosslands, and even southwards to Wales, nesting in the heather and laying about seven or eight light cream-coloured eggs. The cry is so well represented by the syllable "whew" that this word is used as the name of the bird by fishermen; the food consists mainly, it appears, of herbage, and certainly the winter flocks, which begin to feed towards dusk, live chiefly on grass-wrack (*Zostera*) and other aquatic vegetation. The foreign range extends from north of the Arctic Circle in Europe and Asia to about the latitude of Holland, north Germany, and Mongolia; in winter the bird visits North America and other countries, but America has a Wigeon of its own. The Mallard, Teal, and Wigeon are our chief ducks for the market, sea-ducks and diving ducks being mostly uneatable.

SUBFAMILY **Fuligulinæ**, OR SEA-DUCKS

The Pochard or Dunbird (*Nyroca ferina*) has certainly increased as a breeding species in Britain of late years, perhaps, as is doubtless the case with many birds, in consequence of the check imposed by gun licences. But the increase is nothing compared with that of the Tufted Duck, with which the dull-coloured female may be confounded when flushed. She is almost brown, except for the white chin and grey wing-bar, which, it may be remarked, does not shew very distinctly in the air. The male has a fine chestnut head and neck and a mantle pencilled with black and greyish white, his breast and upper back being black. Abroad this species occupies most of temperate Europe and Asia, while in Britain it breeds from the Orkneys to the south of England and in Ireland. For nesting purposes, however, it requires goodly pieces of water, for it builds its nest as a rule well out from the shore among the water plants; broad ditches in marshy ground or small ponds may be occasionally chosen, but invariably near lakes or swamps. The note is low and whistling, with a harsh alarm-cry; the food is mainly of aquatic plants obtained by diving, but when the birds resort late in the year to the sea—though they are by no means regular sea-ducks—the diet is varied by crustaceans and mollusks and the flesh is rank. The nest and. eggs are noticed under the next species.

The male Tufted Duck (*N. fuligula*) is black, with a fine black crest, a white breast and wing-bar; the female is blackish brown, barred with grey below. It ranges from the Arctic Circle to mid-Europe and Asia, being therefore a rather more northerly species than the Pochard. In Britain it has increased enormously during the last half century and breeds from Shetland to the British Channel, where lakes or even large ponds are available. It is still somewhat local, but the present writer once found over fifty nests on one lake. The fabric is of grass, rather than of reeds or sedge, which the Pochard prefers, and the site is near or on dry land, but the eggs, of a dull green colour and often up to a dozen in number, are hardly distinguishable from those of the Pochard, though they are generally elongated and less broad. The Tufted Duck has a guttural voice, and feeds in the same way as the last-named species. Large numbers of both visit us for the winter.

The Scaup Duck (*N. marila*) also comes to us in large flocks, but they stay on the mud flats and rocky shores to a much greater extent than their congeners. In fact it is only since the beginning of the century that the bird has been proved to breed in Britain, both on the mainland of Scotland and in the Orkneys and Hebrides. Abroad its distribution is over Arctic Europe, Asia, and America, but it comparatively seldom nests south of the Baltic. In Iceland it is very common. The harsh cry is said to resemble the word " scaup," but the bird's name is possibly derived from the mussel " scalps " or " scaups " where it regularly feeds, on a diet similar to that of other sea-faring ducks. The nest of rough herbage is placed in thick cover near water or among boulders; the eggs resemble those of the Pochard, but are hardly so green. The whole head and neck is black in the male, rufous in the female; in both sexes the belly is white, the wing-bar white surrounded by black. But in the drake the back is marked with crescent-shaped black lines on a white ground, whereas in the duck it is grey and brown. Moreover she has a broad white band round the base of the bill.

The Golden-eye (*Glaucion clangula*), which ranges from Scandinavia, Russia and north Germany, within the limits of tree growth, westward to northern America and certainly eastward to Mongolia, comes to Britain about October and penetrates far up our larger rivers and to the neighbouring lakes. Abundant though it is from that month onward, and though it occasionally remains till May, this species has never been known to breed with us. It nests in hollow trees—often using old Woodpecker's holes—or in boxes prepared for it by the inhabitants of the north, and lays on a bed of down some dozen eggs of a light but bright green colour. The flight is noisy, the note rather harsh, the food that usual in sea-ducks. The name Golden-eye refers to the colour of the iris, but some of its nearest allies have it of the same tint, though in the Pochard it is red. The general colour of the male is black, with white neck, lower parts, large wing-patch, and shoulder-feathers; the female has the black parts brown, except on the wing, and lacks the distinct white patch on the cheeks of the male.

The Long-tailed Duck (*Clangula hyemalis*) is a decidedly Arctic species found all round the Polar regions of the north, which breeds moreover in the Færoes and Iceland and probably does so occasionally in Shetland. It visits us in winter and remains till late in spring, seldom being very numerous, and less so on the west and south coasts than on the east and north. The drake is a pied bird, beautifully variegated with black and white, and has a fine long pointed tail and a pink middle to the beak. The duck is nearly brown, with white lower parts and no elongated tail-feathers. The curious loud note is thought by the Shetland folk to sound like a repetition of the word " calloo " ; the food appears to consist more of insects and crustaceans than of plants ; the flight is strong and direct. The nest, placed among rough vegetation, consists of little but down, the eggs are rather long and of an unusual greyish green colour. In America this bird is known as the " Old Squaw."

Those who visit the islands of north Britain are almost sure to have met with that nice old bird the Eider Duck (*Somateria mollissima*), if we may be allowed to use the term. When protected, as on the Farne Islands off the Northumberland coast, the sitting duck is so tame that she may often actually be handled, and when she has hatched her young, they may all be seen comfortably swimming close to the sea-weed-covered reef on which the observer is walking. The above-mentioned locality is the bird's only breeding station in England, with the exception of the links near Holy Island, but it occurs on both sides of Scotland up to Shetland and strays to Ireland and southern England in the cold season. The flight is heavy and the voice very harsh ; the food consists of mollusks, crustaceans, and sea-weed, obtained from salt water, for the birds are seldom driven inland even as far as the mud flats. The nest of rough herbage and sea-weed, lined with a quantity of the well-known down, is placed in coarse grass, nettles, or heather pretty close to the sea, or even in the adjacent woods, and contains half a dozen or more large greyish green eggs, often with oily patches of a darker colour. The male is a splendid black and white bird with green nape and neck-patches, a buff breast, and curly white secondaries. The female is plain brown and buff. The Eider Duck is protected on account of its down in Norway, Iceland, and the Færoes, and elsewhere has an Arctic distribution in Europe, western Asia, and eastern America.

The Common Scoter (*Œdemia nigra*), a black bird with an orange ridge on the upper mandible, which has a knob at the base, is extraordinarily plentiful round our coasts from autumn to spring, though gunners have to seek it on the open sea or in wide estuaries. From Iceland it ranges to the Taimyr Peninsula in Asia, by way of north Scandinavia and Russia, while it breeds in some numbers in the far north of the Scottish mainland, and is known to do so in the island of Tiree and north Ireland. It utters harsh reiterated notes, flies at a moderate pace, and feeds chiefly on mollusks. The nest is among grass or heather near fresh water, usually some miles inland; the creamy white eggs are in number from six upwards. The female is brown, with whitish chin and no knob or orange colour on the bill.

The Velvet Scoter (*Œ. fusca*) is not uncommonly met with among the Common Scoters round Britain; abroad it breeds on the lakes of Scandinavia, north— or even south—Russia, and extends to the Yenisei in Asia, but it has not been proved to nest in north

Britain. The drake is easily distinguished by the white patch behind the eye and the white wing-bar; the duck has these marks of a duller colour and no yellow on the bill, though there is a small knob on it. The Surf Scoter (*Œ. perspicillata*), which strays to Europe from North America, visits the British coasts in very small numbers, but perhaps more regularly than is supposed. It cannot be mistaken for either of its congeners, for, though it is also black, it has no white wing-patch, but one on the forehead and another that is larger on the nape, while the orange bill has a rectangular black mark on each side. The general habits of these two species resemble those of the Common Scoter, but the eggs are whiter.

SUBFAMILY **Merginæ**, OR MERGANSERS

The last three of our ducks have long thin bills, with hooked tips and serrated (saw-toothed) mandibles. The male of the Goosander (*Mergus merganser*) is a fine black and white bird, which has a grey rump-region and a salmon-coloured breast, the bill and feet being red. It breeds in the Scottish highlands, where it seems to be increasing, and often is not uncommon on our lakes and larger rivers from autumn to spring. Abroad it ranges from Iceland to Kamtschatka and is known to nest south to Switzerland, while it is, strictly speaking, a bird of the hilly regions in summer. In winter all our species of this group migrate far to the south. Both sexes have a crest, but the head of the female is red-brown instead of black, while she is brownish grey above and duller in other respects. The diet consists mainly of fish, and the voice is harsh. The nest is built by preference in a hollow tree, but may be in a hole in broken ground or in the artificial boxes used by the natives of the north ; the dozen or more eggs are large and of a distinct cream-colour.

The Red-breasted Merganser (*M. serrator*) prefers the sea to inland waters in the cold season, and is much commoner than the last-named " Sawbill," breeding

Red-breasted Merganser

freely in the Shetlands, the Orkneys, and the north and north-west of Ireland and Scotland, both inland and on the coast. The foreign range is north of the Baltic in Europe and similar latitudes in Asia and America. As in the case of the Goosander, the nest consists of little but down, but this is brownish grey instead of white, and the sites chosen are rabbits' burrows, cavities among rocks, and patches of coarse herbage or other thick cover. Moreover, the eggs are greenish buff—less green when incubated—and are seldom as many as a dozen. Otherwise the habits of the two species are similar, and the flesh is equally uneatable. The present bird differs in coloration, for it has a patch of white feathers margined with black on each side near the breast, which is ruddy brown streaked with black, and less white on the wing, while the crest is long and hair-like, especially in the male. The brown female has the head red-brown instead of black and a more distinct black bar on the white wing-patch.

The Smew (*Mergellus albellus*) occurs pretty regularly on our coasts in winter, but more commonly on the east than on the west; the females and young penetrate to considerable distances inland, but the males mostly keep out at sea. This species is now known to breed from Lapland east to Bering Sea; it builds in hollow trees or native boxes and lays about ten rather small cream-coloured eggs on a mass of white down. The bird itself is small for a "Sawbill," with a fine white crest, and black and white plumage elsewhere. The female is duller, while her head is red-brown with no crest.

ORDER IX. COLUMBÆ

In Pigeons and Doves the bill varies in strength, but is always more or less swollen and hardened, with a soft membrane or "cere" at its base; the feet are short, and may be feathered, as in several domesticated races; the toes are without webs, and all on one level, with moderate claws. The wings are long, but rounded; the tail is fairly long; the crop is bigger than in any other Order. The young are hatched naked and helpless, but grow scattered hairs before the feathers. The strong flight, so well instanced in "homing" Pigeons, the reiterated cooing notes, the frail nest of sticks, and the two white eggs are very characteristic.

Family COLUMBIDÆ, or Pigeons and Doves

A wide gap divides the Ducks from the Pigeons, but the latter shew considerable affinity to the Sand-grouse and so to the game-birds. We possess three resident species of Pigeons and one that is truly migratory, the Turtle-dove, which is much more common in some summers than others. The Wood-pigeon, or Ring-dove (*Columba palumbus*), is, however, not only resident but partially migratory, and its two congeners are also partial migrants, immense flocks arriving on our eastern coasts in autumn and soon spreading over all the country, where they do an immense amount of damage to green crops, including turnip bulbs. They appear to come to us chiefly from the direction of Scandinavia, and it should be noted that this species only ranges eastward from that country to about Persia, though it breeds throughout Europe, in north Africa, Madeira, and the Azores. It does not reach Iceland, the Færoes, or Shetland in summer. The bird's strong rapid flight, its cooing note, its habit of clapping the wings over the back when excited, and its frail stick nest with two shining white eggs need no comment, but it may not be so well known that the food is of the most varied description, though always of a vegetable nature. Grain, peas, beans, leaves and bulbs of turnips, and seeds may be mentioned in particular. Breeding takes place early and late, three broods being not uncommonly reared in a season. The colours are compared with those of the next two species under the head of the Rock-dove.

The Stock-dove (*C. œnas*), often shot and sold in shops as the "Blue Rock," has spread enormously in Britain during the last thirty or forty years. It was hardly known to breed as far north as the Border country before 1875, where the first nest was located in Scotland in 1877, but now it is locally common to the extreme north of the mainland and in Ireland. In wooded districts the bird breeds in hollow trees, but it often prefers rabbits' holes on warrens or sea-side links, though trees may be plentiful close at hand. There is generally some attempt at a nest, while the two eggs are of a somewhat yellower white than those of the Ring-dove. The voice is softer than that of its congener, the flight is lighter, the food is the same, as is the foreign range, except that it reaches to central Asia and does not include the Atlantic Islands.

The Rock-dove (*C. livia*) certainly breeds on the western shores of England and Wales, in Ireland, the north and west of Scotland, and especially the islands up to Shetland; but whether our southern and eastern birds, even at Flamborough and St Abb's Head, are not escaped dovecot Pigeons must always remain somewhat uncertain. Those found inland are undoubtedly so. Abroad, however, this species occurs locally in the interior of Europe as well as on the coasts, breeding from the Færoes, Scandinavia, and other northern countries to the Mediterranean basin and the Canaries. The range also extends over western Asia and north India, but complications arise from the presence of nearly allied species in those regions. A bird of the cliffs and caves, where it makes its nest, the Rock-dove passes a little inland to feed, its general habits being those of the Stock-dove. Our

Rock-Dove

three Pigeons are easily distinguished; all are greyish blue with green and purple reflexions on the neck, but the Wood-pigeon and Stock-dove have a much more wine-coloured breast, and the former of these has white on the wings and sides of the neck. The latter has broken bars of black on the wing and shews no white anywhere; the Rock-dove has a white rump and two distinct black wing-bars. All have a blackish tip to the tail; in all the sexes are similar.

The Turtle-dove (*Streptopelia turtur*) has been gradually extending its range in our islands to the northward, and now breeds sparingly up to the Borderland. It is very abundant in our eastern counties and fairly plentiful on the west, arriving in very late April or May and leaving about September. It flies quickly for moderate distances, has a coo which sounds in the distance like the purring of a cat, and feeds upon grain, peas, seeds of weeds, and the like. After breeding it flocks to the stubbles. The nest is very slight and generally to be found in hedges, bushes, or low trees; the eggs are often more pointed than is usual in the Family and slightly cream-coloured. The bird itself is in both sexes brown with greyer head, neck, and rump; the lower parts are wine-coloured, the tip of the tail, very conspicuous in flight, is white, and there is a patch of black feathers tipped with white on each side of the neck. On migration the Turtle-dove has even occurred in Shetland and Lapland, but it normally breeds from about the Baltic to Turkestan as well as to the south of Europe, Madeira, the Canaries and Egypt. Its range overlaps that of other similar species, such as the Asiatic Turtle-dove, well known to us as a cage-bird.

ORDER X. PTEROCLETES

The desert-loving forms included in this Order were once thought to be very closely akin to Grouse, and are certainly connected with them, but it now appears that they come nearer to the Pigeons and the Plover tribe. They have a short arched bill and short feet, with toes which, in the only British species, are almost joined and cased with hairy feathering, the hind-toe being here absent; long pointed wings and long tail, the outer wing-feather on each side and the two central tail-feathers being elongated into thin points; a large crop; and downy young, which run from the shell. They walk well, and fly swiftly after the manner of Golden Plover, while it is a remarkable fact that an Asiatic species alone visits us, and not one of its southern allies.

Family PTEROCLIDÆ, or Sand-grouse

To all lovers of birds Pallas's Sand-grouse (*Syrrhaptes paradoxus*) affords one of the most curious cases of irregular migration. Usually it moves only to short distances from its summer quarters on the barren steppes of central Asia, but for some unknown reason, at intervals of many years, it makes incursions in huge numbers to western Europe. Small irruptions took place in 1859, 1872, and 1876, but the main instances were in 1863 and 1888. On both of these occasions the birds reached us in May, and on the second a special law was passed to protect them; but they all disappeared, with a few exceptions, before the following spring. No eggs were obtained nearer than Holland and Denmark in 1863, but in 1888 two sets were found in Yorkshire, while a nestling was picked up on the Culbin Sands near Nairn in Scotland by a keeper, who discovered another in 1889. This incursion finally reached the extreme west of Ireland. The flight of this Sand-grouse resembles that of the Golden Plover, being fast but often circling; the birds customarily scrape out holes in which they sit like barndoor fowls, and then rise unwillingly; sandy districts naturally attract them most. The note is of a clucking nature, the food of various seeds and insects. The three eggs, deposited without any nest, are yellowish buff with purplish brown spots and blotches. The male is a black and buff bird, with grey head, wing-quills, and tail, the central feathers of the last being greatly prolonged; the throat is rust-coloured, while the breast and belly are transversely marked with black. The female has the head like the back and a black stripe across the throat.

ORDER XI. GALLINÆ

We have now arrived at the very large group which contains the Grouse, Pheasant and Partridge alliance, besides our domestic fowls and many splendid foreign birds, of which the Peacock is best known to the multitude. The head is comparatively small, with a short stout curved bill; the feet are powerful and vary in length, but in British forms the hind-toe is always elevated; the front toes are feathered in the Ptarmigan and Red Grouse, as is the whole foot in Grouse generally. Spurs, from one to three pairs, are not uncommon; combs, wattles and bare skin on the cheeks or round or above the eye are prominent features of many species. In Grouse the bare red skin round the eye is pimply. The wings are generally rounded; the tail varies greatly, as may be seen in the case of Pheasants, Partridges and Fowls, while it is called vaulted in the last-named. All the members of the Order possess a crop, and the downy young run at once and soon learn to fly.

Family TETRAONIDÆ, or Grouse

The Sand-grouse is, as we have seen, a Grouse only in name, but we have four representatives of the

Family in this country, the Capercaillie or Wood-grouse, the Black Grouse, the Red Grouse, and the Ptarmigan. The Capercaillie (*Tetrao urogallus*) used to breed in ancient times in all four divisions of the realm, but finally became extinct in Scotland before the year 1800. In 1837, however, it was reintroduced to the Loch Tay district, and is now common in many parts of central

Hen Capercaillie

Scotland, and has spread past the Grampians to Ross-shire as well as southward to Wigtonshire. Scandinavia and Russia are its great strongholds, but it is found on the great southern ranges of Europe and across Siberia to Lake Baikal. This magnificent game-bird, which seems even bigger than it is when it falls to the gun, is, in the male, blackish grey above and pure black below with green-glossed chest; it has a heavy and deceptive flight, much swifter than it appears, though when the young have just left the nest the cock is content to move from branch to branch near the ground and act in company with the hen as guardian. The six or more eggs are deposited in a hollow scraped in the ground under shelter of a bush or log or at the base of a tree; they are yellowish with small close-set orange markings or blotches. The food consists mainly of Scotch-fir shoots and berries, and in part of worms and insects, so that conifer woods are necessary to the bird's well-being. The cock is polygamous and in spring performs antics before the hen, which Scandinavians call his " lek " or " spel "; these he accompanies by hoarse noises slightly reminiscent of similar sounds made by Pigeons. The hen is mottled with brown, buff, and white, the breast being of a brighter buff.

The Black Grouse (*Lyrurus tetrix*) may be shortly passed over, as the habits are much as in the last species. The colour, however, is black throughout, except for the white wing-bar and coverts under the widely forked tail, the female being red-brown with black markings. This species eschews thick woods, but loves bracken-covered slopes and rough damp country of many descriptions; it eats insects, berries, seeds, and buds of various plants and does not restrict itself to fir-shoots; at harvest-time it cannot be kept away from the corn "stooks" in the fields, if any be near. In spring it utters a softer note than the Woodgrouse, and when sitting in a tree, as it often does, gives a still better imitation of a Pigeon's coo. The scanty nest is in heather or rough herbage; the eggs are similar to those of the Capercaillie, but smaller. Hybrids of the two are occasionally found. Our form is said to be peculiar to Britain. The typical form occurs only in Europe.

The Red Grouse (*Lagopus scoticus*), our familiar red-brown game-bird, which is not uncommonly much

Red Grouse

blacker or more varied with white than the most typical specimens, is what is termed "endemic," that is to say confined to this country. Its moorland haunts are known to us all, and in its choice of a residence it thus differs from the Willow Grouse of the north of the Old World, which is more partial to scrubby localities. Yet it sometimes perches on trees, as that bird constantly does, while many persons consider them as merely different races of one and the same species. The flight is very powerful, as sportsmen know to their cost, and after breeding Grouse are accustomed to gather in packs; the food consists mainly of the tips or tender shoots of ling (*Calluna*), which must therefore be regularly burnt to ensure a constant supply of young growth. Otherwise moors become " diseased." The diet is varied by berries and seeds, especially, it appears, those of the moor-rush (*Juncus squarrosus*). The alarm-note of the male, " cok, cok," his " crow," and the hoarser call of the female are matters of common knowledge to those who visit our moors in Scotland, Ireland, Wales, England north of Derbyshire, Shropshire, and Staffordshire. In Shetland, Surrey, Norfolk, and Suffolk the bird has been introduced, but with little success except in Suffolk near Mildenhall. Abroad the same may be said of south Sweden and the west German frontier. The nest, nearly always built in heather, is of the slightest description; the eight or more eggs are yellowish with blackish, reddish, or purplish markings, and are often very beautiful. In the Red Grouse, the hen-bird is smaller and distinctly yellower, but in its congener the Ptarmigan (*L. mutus*) the sexes are very different. The cock in summer is grey and brown with black lores and tail, the tip of the latter and the belly being white. The hen is reddish buff with black markings. But the wings in both are white, and the birds turn entirely white in winter, except for the tail and lores. The Willow Grouse does the same, but has even the lores white. Moreover the Ptarmigan has a third or autumn dress. In south-west Scotland, and possibly in the English Lake district, this species used once to be found; now it is confined to the higher Scottish mountains, for it is always an inhabitant of stony hill-tops, which it only leaves for somewhat lower altitudes in time of snow. On these stony slopes or flats it lays eight or more eggs, generally redder than those of its congener, with little if any nest, and there the birds pack after breeding. In similar spots the Ptarmigan is found from Scandinavia to the Urals, as well as in the Pyrenees, Alps, and the mountains of Austria; but in Spitsbergen a larger species takes its place, while in Iceland, Greenland, and at lower altitudes in Arctic America and north Asia still other species occur. The croaking cry of the cock, as he keeps watch near the sitting hen, is often the only bird's note heard on the desolate wastes of stones; the flight is fairly strong and fast, and the food of the same nature as in the Red Grouse, with a less proportion of heather-shoots.

Family PHASIANIDÆ, or Pheasants, Partridges and Quails

We may spare ourselves the description of the Pheasant (*Phasianus colchicus*), which is to be seen in every game-dealer's shop, but content ourselves with remarking that the white collar, seldom absent from the cock, is due to a cross with the Chinese ring-necked species, and with calling attention to his brilliant plumage as opposed to the brownish and comparatively short-tailed hen. Many other kinds of Pheasants have recently been introduced to Britain, but the typical bird dates back to about the time of William the

Conqueror, if not to an earlier period, when it is supposed to have been brought from the river Phasis, east of the Black Sea. Thence it only ranges eastward to the Caspian and westward to Greece and Albania, though many other species are found in the river valleys of Asia, for Pheasants are naturally inclined to damp or swampy localities. It is probably native

Pheasant

in south-east Europe. In Britain our birds occupy the majority of the wooded districts, though less plentiful in the north. The extremely rapid flight and quick rise from the ground are characteristic, the crow of the cock is heard throughout the land, and the dozen or more olive eggs are probably known to most of us. The nest is flat and scanty. The food, besides artificial diet, consists of grain from stubbles, seeds, berries, leaves, snails, insects and their larvæ, so that from an economic point of view the Pheasant is as beneficial to the farmer as it is important to those who provide our food supplies. The same may be said of that even commoner game-bird the Partridge (*Perdix perdix*), which is native, except where it has been introduced in the northern islands. It pairs very early, and any low cover, thick or scanty, serves to hold the flimsy nest, the eggs being much smaller and lighter in colour than those of the Pheasant, and often more in number. The coveys, which "pack" into larger companies towards autumn, the voice, the flight, and the methods of shooting the bird call for no special notice here. The food resembles that of the Pheasant. Rare in Scandinavia, the Partridge is plentiful thence to central and eastern Europe, the Tian Shan and Altai Mountains, while it occurs down to north Spain and mid-Italy. All are familiar with its coloration, but the sexes are troublesome to determine unless adult; a safe guide, however, is to be found in the wing-coverts, which have buff cross-bars in the female and longitudinal stripes in the male.

The Red-legged or "French" Partridge (*Caccabis rufa*) is now well established in eastern and southern England, but does not seem inclined to spread to the north, west, or even south-west. It was acclimatized in our country about 1770, and is found in France, Belgium, Portugal, Switzerland, Spain, the Atlantic and Balearic Islands, Elba, Corsica, and north-west Italy. It will be wise, however, not to fix any precise limits, as the African Barbary Partridge (*C. petrosa*) and the more eastern *C. saxatilis* no doubt overlap the range of the Red-legged Partridge in places. The flight is exceptionally strong and direct, and the note more grating than that of the Common Partridge, but the diet is similar. The eggs are larger and of a pale

Red-legged Partridge

yellowish colour with faint rufous spotting. As the bird is a strong runner, sportsmen used to dislike its presence, but in these days of driving with beaters the objection has practically vanished, and the fable of its harassing its kin is now but seldom heard. The upper parts are bright brown, with grey crown and chestnut tail; the throat is white, the breast grey, the flanks grey beautifully barred with chestnut and black, the belly buff; a black gorget surrounds the throat; the bill and feet are red.

In some seasons the Quail (*Coturnix coturnix*) is much more abundant than in others, though, possibly owing to the immense numbers killed on passage in the south of Europe, it no longer breeds with us regularly. Nests have, however, been found even in Shetland, and the bird ranges from the Færoes across Europe and Asia, as well as to north Africa. In the rest of Africa, the Atlantic Islands, Japan, and elsewhere we meet with different forms, perhaps hardly to be considered distinct species. When Quails were plentiful in Britain, only a small proportion of them were resident. In shape like a Partridge the male is of a similar brown colour, but instead of a brown "horse-shoe" on the breast has two brown bands on the sides of the neck running down to a patch of a blacker tint in front. The colours in the female are less distinct, but she is bigger than the male, as is the case in the allied foreign genus *Turnix* (Button-quail). The habits agree with those of the Partridge, though the nest, of the slightest description, is more often among growing crops, while the eggs are whitish boldly marked with brown. The triple note of the male is very peculiar and can hardly be mistaken; that of the female is low and attracts little attention.

Order XII. Grallæ 179

ORDER XII. GRALLÆ
Suborder Fulicariæ
Family RALLIDÆ, or Rails, Water-hens, and Coots

The members of this Suborder are skulking marsh-birds, which have downy young that swim at once. The bill is not very long and in Water-hens and Coots develops into a naked plate or "shield" on the forehead; the feet are long as a rule, while the lengthy toes have a membranous edge in Water-hens, and membranous lobes in Coots. The wings and tail are comparatively short.

The Land-rail or Corn-crake (*Crex crex*), which visits us in numbers between April and September, and occasionally remains later, has for some unknown reason decreased considerably in our eastern districts during the last thirty years, though perhaps not elsewhere. Fields of growing corn, grass, and clover, rush-beds, and thickets of low gorse are its favourite haunts, and there it lays some seven or eight dull cream-coloured eggs with rusty red and lilac markings in a hole scraped in the soil, with a little lining. The Land-rail is a ventriloquist and it is often difficult to say whence its craking voice proceeds, while both sexes are equally hard to flush; the flight is slow and heavy, the food more of worms, slugs, and insects than of plants and seeds. This species, which is brown, lighter below and with darker markings above, ranges over northern and central Europe and Asia and migrates far south.

The Water-rail (*Rallus aquaticus*) is a skulking and silent species known to breed in many marshy localities in England, southern Scotland, and Ireland, but specially in the sedge-beds of the Fens and Broads. It is with us all the year, though its numbers are increased from abroad in the cold season ; in summer it ranges from the Arctic Circle in Europe and Asia to north Africa and Turkestan. It is a darker bird than the Land-rail and of a more olive colour, the cheeks and under parts are bluish grey, the bill is red, and the

Water-rail on nest

flanks are heavily barred with black and white. The food is of the same nature as that of the last species, allowing for the aquatic habits, but the flight is more feeble, while the spring cry resembles a groan, and is called in Norfolk " sharming." The nest, placed in very thick sedge, generally where there is standing water, is made of the broad leaves of water plants—

Grallæ 181

reeds and so forth—and contains about eight creamy white eggs with lilac and red-brown spots and streaks.

The Spotted Crake (*Porzana porzana*) migrates to us in March and generally leaves in the late autumn. It used to breed commonly in East Anglia and sparingly in the bogs of other parts of the kingdom, including Ireland, Wales, and Scotland below the Caledonian Canal, but now the discovery of a nest must be an exceptional occurrence. The structure is smaller than that of the Water-rail, but is built in similar places; the spots on the eggs are darker and by no means so large, giving them an entirely different appearance. The spring note is much clearer and less mournful; otherwise the habits of the two species are almost identical. The male is brighter than the female, and is olive-brown above with small white spots, the face and throat being grey, the lower parts lighter with broad dark stripes on the flanks. This Crake breeds from mid-Scandinavia and Russia to the Mediterranean southwards and west Asia eastwards.

The Little Crake (*P. parva*) belongs to the list of irregular visitors, but must be mentioned here as having been mistaken for Baillon's Crake (*P. pusilla*), though it has practically no white spots above. Both are very small as compared with our other Rails, and only visit us in spring and autumn. Baillon's Crake, however, has remained at least four times to breed in the eastern counties, which is not astonishing, as it does so regularly in Holland, Russia, France, Italy, and thence to South Africa. Instances of the bird's occurrence are very rare in Scotland and Ireland, and they are hardly more frequent in England, no nests having been found since 1858 in Cambridgeshire and 1866 in Norfolk. The

182   Order XII

habits are even more retiring than those of the Spotted Crake; the flight, food, and cry are not dissimilar; but the nest is generally found near smaller pieces of water, and the eggs resemble those of the Little Crake in being yellowish brown with dark brown blotches and streaks. The colour of the plumage above is brown with small black and white spots, the cheeks and lower

Coot

parts are blue-grey, the flanks barred with black and white.

The Moor-hen or Water-hen (*Gallinula chloropus*), though it appears black, is really very dark brown with equally dark grey lower surface; the yellow bill backed by a red plate on the forehead, the greenish legs with their red "garters," and the white under tail-coverts being other noticeable features. Except in the Arctic regions, it breeds throughout Britain, Continental Europe, Asia, and Africa, very closely allied species occupying America, Australia, and certain islands. The note is of a craking description, and is most commonly heard towards dusk ; the flight is low and laboured ; the food consists of insects and their

Coot's nest and eggs

larvæ, worms, herbage, or grain in winter. These birds, naturally shy, come out boldly to feed on grass-land after rain or towards evening. The nest is a big flat mass of the leaves of sedge, reed-mace, and other plants ; the eight or nine large eggs are stone-coloured with reddish or purplish spots ; the structure is placed in various low situations, generally in marsh or river vegetation, sometimes on tree-stumps near water, less commonly in thick old hedge-rows over ditches.

The Coot (*Fulica atra*) is blacker and larger than the Water-hen with merely a white stripe on the wing and a large white plate on the forehead. As is the case with the latter, it breeds from Shetland to the Channel, but requires larger pieces of water and is therefore more local. Abroad it ranges over Europe and temperate Asia to the Azores, north Africa, northern India, the Philippines, China and Japan. The flight is comparatively strong though low, the note an unmistakeable croaking sound, the food similar to that of the last species; but the nest, nearly always built in water some feet deep, is much larger and more solid, and the eggs shew small blackish specks on a yellowish ground.

Suborder Grues
Family GRUIDÆ, or Cranes

The Cranes, sometimes confounded with Storks, are tall birds with very long necks and feet, a hind-toe placed very high above the others, and long stout pointed bills. Our species, when adult, has a warty red patch on the head and long drooping secondary wing-feathers. It is entirely a marsh bird in the breeding season.

To students of antiquity the Crane (*Grus grus*) and the Great Bustard (*Otis tarda*) are of exceptional interest, but whereas the latter bred in eastern England until the earlier years of last century, the former ceased to do so before the year 1600. It nested in the marshes of East Anglia, but now is quite an irregular visitor from its homes in Europe and Asia. In both continents the bird is widely distributed, but only breeds in wild

Grallæ 185

districts where large morasses occur, though it does so as far north as Lapland and Finland in Europe; the nest is a large mass of aquatic vegetation, while the two or three big eggs are of a curious greenish grey colour with indistinct reddish brown markings. Cranes fly magnificently, and from olden times have been signs

Common Crane

of the season on their migration to and from the south, when they fly at enormous heights; the food consists of almost any kind of vegetable substance to be found on the ground—roots, fruits, herbage, and the like, with insects and doubtless frogs and small mammals. The note is loud and trumpeting. In both sexes the colour is blue-grey, becoming darker on the wings, and adults have a bare red crown-patch on the black and white head. Herons are often termed Cranes in Scotland by the ignorant, a fact which may cause erroneous records.

Suborder Otides
Family OTIDIDÆ, or Bustards

Bustards have short thick bills, long feet with

Great Bustards

stout toes, moderate wings, and soft rounded tails. The three properly developed toes all point forwards. These birds are large and heavy and inhabit bare plains, sandy wastes or wide corn-fields.

The Great Bustard is as familiar to most people by name as the Great Auk, but with better reason, for it bred in Norfolk and Suffolk till about 1838, on the downs and wolds of other eastern and southern counties earlier in the century, and at least up to the year 1526 in Berwickshire and East Lothian. It seldom visits Britain now, but ranges over central Europe and Asia probably to the boundaries of China, being also found from the Spanish Peninsula to Syria. The male has a grey head, warm buff upper parts marked with black, white on the wings and belly, chestnut and grey breast-bands, and a tuft of long white bristles on each side of the gape. The female is much smaller, with no bristles and no bright colour on the breast. This splendid species has a very strong flight, except during the June moult, at which the males lose all their wing-quills at once; they are pugnacious and shew off like a Turkey in spring-time, when they swell out enormously a pouch below the tongue. They feed upon green crops, varied by seeds, worms, or even small mammals and birds. Bustards afford excellent sport and are good for the table, but they need very careful stalking, unless the gunner lies in wait on their line of flight. The two or three eggs, deposited without any nest, are olive-green with faint reddish blotches. After their spring fights the cocks flock apart from the hens.

The Little Bustard (*O. tetrax*) is a rare visitor to England as well as Scandinavia and north Russia, but its visits to us, though irregular, are constant, as might be expected from its abundance in some districts of France and in the Spanish Peninsula. Thence it extends on both sides of the Mediterranean to Greece, Turkey, south Russia, Turkestan and west Siberia. The flight is noisy, the note monosyllabic and reiterated; the food resembles that of the last species, but the smaller eggs are as a rule much brighter, and the bird makes a slight nest in rough cover. Each cock remains with his chosen hen. The former is buff and black above and chiefly white below, while the neck is banded in curious but beautiful fashion with black and white. The hen lacks these adornments. This species cannot swell out its throat to the same extent as the last-named, but the same fact holds to a less extent.

ORDER XIII. LIMICOLÆ

The members of this very large Order are chiefly marsh or moorland birds. The bill varies much, being quite hard in such forms as the Oyster-catcher, soft with a hard tip in most Plovers, similar but without the hard tip in Sandpipers, and provided with an abundance of nerves in Snipe and Woodcocks. The feet are immensely long in Stilts and Avocets, but are more usually moderate ; the wings vary, the tail is usually short. In the Peewit, for instance, the wings have an enormous spread for a bird of its size. The toes are often only three in number, for example, in the true Plovers (*Charadrius*) and the Sanderling ; in Phalaropes they have membranous lobes, as in Coots. The young, which run from the shell, are generally covered with yellow down marked with longitudinal brown stripes ; but occasionally they are grey or more varied with black, red, orange, or white.

Family ŒDICNEMIDÆ, or Stone-Curlews

On the extensive warrens of Norfolk and Suffolk—and exceptionally on the downs of the southern counties—we meet with that interesting and retiring bird, the "Stone-Curlew" or Norfolk Plover (*Œdicnemus œdicnemus*). It arrives about April, and a few pairs never proceed further inland than the Dungeness

Order XIII. Limicolæ

shingles in Kent, but the majority pass on to East Anglia, though a limited number breed elsewhere on our downs and heath-lands as far north as Yorkshire. Two big stone-coloured eggs with grey and brown spots or scrawls are laid in April or May in a shallow hole scratched in the ground, which is almost always lined with rabbits' dung; the hen-bird will sit very

Stone-Curlew

closely if taken unawares, but usually rises two or three hundred yards ahead of an intruder. The downy young are rather sluggish, and are very difficult to find when they squat on the ground to escape notice. The mournful call of the Stone-Curlew is of the same nature as that of the Golden Plover, and is louder after dusk than in daylight; its food consists mainly of mollusks, worms, and insects; its flight is low and in summer-time short. In fact its habits are not unlike those of other Plovers, except that it is shy and less easy to flush. In colour it is sandy brown above and buff below, with darker markings, the throat and belly alone being whitish; the eyes are large and yellow. This species ranges over central and southern Europe, north Africa with the Canaries, and central and south-western Asia to India, Ceylon, and Burma.

Family CHARADRIIDÆ, or Plovers and their Allies

The Dotterel (*Eudromias morinellus*) used to be well known on the spring and autumn passage in many parts of England, but is now rare, owing perhaps to the cultivation of the wilder localities. Nevertheless individuals, and even small parties, still occasionally halt on their way, and it would be a bold man who would assert that so retiring a bird had entirely ceased to breed on the hills of Lakeland. But it does so no longer in Dumfries-shire and Galloway, though it is scattered over the hill-tops in the Grampians and the Cairngorms, where it lays three eggs, much like those of the Peewit, in a depression in mossy ground with no real lining, unless it be a few bits of lichen. The food is of worms and insects, the note querulous at the nest, while the flight, though at times swift, is very short when the birds move around an intruder with every sign of anxiety. Abroad the Dotterel ranges over the tundras and fells of northern Europe and Asia; it might almost be considered an Arctic species if it had not been found breeding in Bohemia, Styria, Transylvania and the hills of central Asia. The general coloration is light brown, but the crown and belly are black, as are the tips of the feathers above a white chest-band which precedes the fine chestnut breast. There is also a distinct white band from each eye to the nape, while the throat and under tail-coverts are white also.

One of the most familiar of our coast species, though there confined to sandy and pebbly shores while extending to sandy inland warrens and lake-sides, is

Ringed Plover

the beautiful little Ringed Plover or "Sand-lark" (*Ægialitis hiaticula*), buffish brown above and white below, the head and neck being varied with black and white, and the latter colour forming a broad collar. Generally it is a tame but wary bird which paddles about in shallow water or patters along for some distance before it rises; its flights are short, quick, and often circuitous; its note is clear and ringing; its food consists of the worms, insects, crustaceans, and so forth which it can find in the damp sand or ooze. This species breeds in the Arctic regions from eastern America, Iceland and Greenland to Lake Baikal, but is more common southwards; a smaller form—which, however, is not the "Little Ringed Plover"—ranges to the Atlantic Islands, north Africa, and Turkestan. A hole is scraped in the ground to contain the four pointed buff black-spotted eggs, which, as is the custom in the family, are placed with the pointed ends inwards; on the East Anglian warrens this hole is lined with small stones and gives the bird the name of Stonehatch.

The Kentish Plover (*Æ. alexandrina*) is similarly coloured, but has a broken instead of an entirely black chest-band and much less black on the head; an illustration or a skin should here be examined. It is a migrant to Britain, arriving in April and departing in mid-autumn, while it only breeds on the shingly shores of Kent, Sussex, and in the Channel Islands. This Plover ranges from the Baltic southward in Europe, and reaches the Atlantic Islands, north Africa, and central Asia as well as China and Japan, but requires lakes or sea-coasts for its breeding quarters. The unlined hole which contains the eggs is usually in fine or shelly sand, the eggs themselves being rougher than those of the Ringed Plover, duller in ground-colour, and scrawled rather than spotted with black. The Kentish Plover is a smaller and less obtrusive bird than its congener, with a feebler cry; otherwise the habits are identical.

The Golden Plover (*Charadrius apricarius*) breeds on the moors, generally at a considerable altitude, from Shetland to Derbyshire, in Wales and its border counties, in Somerset, Devon, and Ireland; in autumn it comes down to the low country and during winter is common on many parts of the coasts, as well as on the flat fields of East Anglia and other districts. The upper parts are in summer black closely spotted with yellow, the lower parts uniform black; from autumn to spring the latter are brownish white. The forehead and a line running from above each eye to the flanks are always white. From August till late spring many immigrants from abroad are with us, the foreign range extending over Arctic and northern Europe and Asia to the Yenisei. A few pairs even breed in Switzerland. The food consists of insects and their larvæ, worms, and the like, with small mollusks from the shore, and at times a little vegetable matter; the note, very characteristic of our wilder moorlands, is a sharp continually repeated whistle; the flight, especially of the winter flocks, is low, swift, and often circling; the nest is little more than a depression in the grass, heather, or moss, and contains three or four yellowish buff eggs with fine blackish blotches.

The Grey Plover (*Squatarola squatarola*) can be distinguished with certainty from the last species by its black axillaries, as these feathers below the wing are white in the Golden Plover; the spotting, moreover, is white rather than yellow, and there is a distinct hind-toe. The Grey Plover only visits our shores from autumn to spring, and very rarely occurs inland; wide stretches of sand and mud flats are its ordinary haunts, while it is more common in the east than in the west. Its summer range extends over the Arctic portions of Russia, Asia, and America; the habits, nests, and eggs differ but little from those of its congeners.

It is unnecessary to describe in detail the familiar Lapwing or Peewit (*Vanellus vanellus*), with its general metallic black and white coloration, its buff tail-coverts,

Lapwing

and its long pointed crest. The bird is common in every part of our isles and more or less resident, though its numbers are greatly augmented in winter by immigrants from the north. The slow flapping flight of its broad rounded wings is very characteristic, and still more so its habit of wheeling round an intruder on its breeding quarters. These are as often on moors as on grass-fields or ploughed land, while the nest of a few straws or grass stems is sometimes conspicuous by its absence ; the well-known eggs, sold in quantities for eating, are laid from March to June, and are stone-coloured or greenish with blackish brown markings. The present writer found four in 1913 at an altitude of 3000 feet in Scotland. The food consists of insects and their larvæ, worms, slugs, and small creatures from the beach, for Lapwings resort in numbers to the shore at low tide. Flocks are formed after the breeding season and are very constant to the parts of the country where they first congregate in autumn; the same flock may even be seen through the winter till spring, occupying the same two or three fields. The note of " pee-wit " or " pease-weep " has given the bird more than one of its names. South of the Arctic Circle it ranges over Europe and northern Asia and even reaches northern Africa, though far less common in the south.

The Turnstone (*Arenaria interpres*) ranges over the Arctic regions of both the New and the Old Worlds, and visits us between July and May, or more rarely remains till June. It has not, however, been proved to nest south of the Baltic. With us it frequents rocky shores where sea-weed is plentiful; in the far north it retires to the fells to breed. The locality most usually chosen is, however, some spot near high-tide mark, where the four greyish green eggs, with grey and brown spotting, are deposited with scarcely any bedding. The food of small mollusks, crustaceans, and insects is sought among sea-weed and stones, which, as the bird's name implies, are often overturned in the search. The note is of a whistling nature, the flight quick but not long sustained. The mantle in the Turnstone is chestnut and black, the lower parts white, while the rest of the plumage is a mixture of black and white. The legs and feet are reddish orange.

The Oyster-catcher or Mussel-picker (*Hæmatopus ostralegus*) is a much larger black and white bird with flesh coloured legs, crimson irides, and orange-red bill.

Oyster-catcher

It frequents our rocky shores and sandy bays, and breeds either among the sand and shingle or actually on the low rocks, as well as in salt-grass or stretches of sea-pink. No nest is made, the eggs, generally three in number, are yellowish buff with black spots or scrawls. The flight is low and strong, the note piercing and reiterated, the food of shell-fish and other sea animals, with a little vegetable matter. In summer this species is much commoner in Scotland than in England and reaches Shetland; in winter many migrants visit us from abroad, where the bird ranges over Europe and central Asia. Even in north Britain many pairs breed inland on the stretches of shingle by the rivers, and the same fact holds good of the

Oyster-catcher's eggs

Continent. The local name "Sea-pie" of course means Sea Magpie.

The Avocet (*Recurvirostra avocetta*) is to us a species of more than usual interest, as it nested in England till about 1824, and might still do so if the individuals and small flocks, which visit us on passage in spring, were unmolested. The bird used to breed in swampy ground from the Humber to Suffolk, in Kent and in Sussex; it now does so locally in temperate Europe and Asia and southwards in Africa to Cape Colony. The note is a clear "whit," the food of insects, worms, crustaceans, and so forth, the flight often performed in circles. The nest, if any, is very slight, the three or four stone-coloured eggs with black blotches often lying on bare mud or sand, where there

Avocets

is little or no vegetation. In colour the Avocet is plain black and white, but it has a most extraordinary bill, which has been likened to a couple of long flat pieces of whale-bone pointed and upcurved at the tips. The feet are also long, and the toes half-webbed. The food is scooped up with a sidelong action of the bill, whence the bird has been called the "Scooper." Other names are "Cobbler's awl" and "Shoe-horn."

Limicolæ 199

The Red-necked Phalarope (*Phalaropus lobatus*) is another exceptionally interesting bird, being one of those confined, in our islands, to the Shetlands, Orkneys, Hebrides, and one locality in Ireland. In colour it is mainly grey above and white below, but is varied with rust-colour and white on the upper parts, and has rich chestnut on the sides and front of the neck, while the breast is grey and the tail brown. In this genus it happens that the male is the smaller and duller bird. During winter the head and upper parts generally are much whiter. This Phalarope only remains in its British breeding quarters from May to August, and very few visit us on migration; but from Iceland and the Færoes to the fells of Scandinavia and the Arctic regions of both worlds it is abundant in summer. A pretty little nest of grass and the like is built in a tuft of herbage in marshy places, and four greenish or brownish eggs with black-brown markings are laid. The parents are very tame. The note is a querulous tweet, the flight wavering, the food of insects, worms, and crustaceans. The bill in this genus is long and straight, while the toes are furnished with lobed membranes, as in the Coots. *Phalaropus* means "Coot-foot."

The Grey Phalarope (*P. fulicarius*), a larger bird with shorter beak, has an Arctic range which extends all round the north polar regions; it migrates to Britain only in certain years, though then often in great numbers to our more southerly coasts, while it has not uncommonly been killed far inland. It is only in winter that this species is grey; in summer it is chestnut below and black and rufous above, the cheeks and a bar on the wing alone being white. The head is nearly black. The habits are those of the last-named bird, for in both cases the male incubates and the female courts him ; the nest, however, is of necessity on barer peaty or mossy ground; the eggs, which are often less pointed, and more olive, are hardly distinguishable.

Everyone knows the Woodcock (*Scolopax rusticula*) by name or picture, and few have not heard of the

Woodcock

regular " flights " that visit us from the north in the middle of autumn. The flocks take their departure in March, until which time the members separate and search for suitable feeding ground. Boggy moors and damp woods containing springs are their favourite resorts, the north and west of Ireland being the most notable winter-quarters, though the west of Scotland,

and particularly the island of Tiree, leave nothing to be desired from the point of view of the shooter. A certain proportion of the birds nest in Britain and seem to be for the most part resident—especially in the north; they lay four large creamy or brownish white eggs with grey and rust-coloured spots in a depression lined with dead leaves, but it is of little use to search for them except in swampy woods, where the ground is covered with dead leaves of such trees as the oak, though heathery and bracken-covered openings also prove attractive. Roughly speaking, the Woodcock ranges from the Arctic Circle over temperate Europe and Asia to the Himalayas and Japan, and to the Atlantic Islands, but many localities are not suitable. The twisting flight of the birds, which lie very closely until flushed, is characteristic; the food resembles that of the rest of the Family, the insects, worms, and so forth being obtained by probing the soft marshy ground with the extremely long and sensitive bill. At dusk and about sunrise the males constantly fly round a fixed " beat," uttering a curious hoarse note, though they have a sharper cry as well; both sexes apparently follow this practice when going to feed, and may be relied upon to return to the same point of their beat, as a rule, in less than half an hour. This habit is called " roading." The female also removes her young from place to place, carrying them between her legs. The upper plumage of the Woodcock is ruddy brown and the lower parts pale brown, both being waved or streaked with darker colour. The large eye is very noticeable.

The zigzag flight of the bird, its alarm-note of " scape-scape " when flushed, and its habit of probing fenny or boggy ground for insect-food are characteristics inseparably connected with the Common Snipe (*Gallinago gallinago*), which is distributed in suitable places throughout the United Kingdom. The erratic flight is a source of great trouble to young shooters, who are neither quick enough to fire at the exact moment that

Common Snipe

a bird rises, nor experienced enough to let it settle to a steady pace. Snipe are very closely connected with Woodcock, are equally long-billed, and are addicted to a similar diet, but they are much smaller and are constantly seen in companies or "wisps," which rise at considerable distances. On the other hand, a single Snipe lying hidden in grass may sometimes be touched before it will stir. The Common Snipe ranges over northern and temperate Europe and Asia, except the far north and the warmer parts of the south, and even reaches north-west Africa; in winter hundreds of immigrants are added to our breeding stock, arriving late in October and leaving us in March. From April onwards those that remain build a small nest of grass among the herbage in swampy places on our moors, fens, and marsh-lands, laying four olive-green or yellow-brown eggs with oblique blotches of grey and brown. These are wonderfully large for a bird of the size. It is at this season that most Snipe "drum," that is, produce a curious noise in descending from aloft, which is caused chiefly or entirely by the vibration of the outer tail-feather. The present writer has, however, constantly had Snipe drumming round him when shooting in winter. The colour of the upper parts is brown with buff streaks and blacker spots; the under parts are white, becoming brown with darker markings on the fore-neck and flanks.

The Great Snipe (*G. media*) is a larger and more boldly marked bird, which is often confounded with the last species, but can always be distinguished by having sixteen instead of fourteen tail-feathers. The Solitary Snipe, as it is sometimes called, is a regular but uncommon visitor to Britain in autumn or even in spring; its home is Scandinavia, Denmark, and Holland, northern Germany and Russia, with temperate Asia as far west as the Altai Mountains. It is rarely seen in flocks.

The Jack Snipe (*Limnocryptes gallinula*), with only twelve tail-feathers, is a much smaller bird, which has erroneously been stated to breed in Britain; its summer range extends from Scandinavia and north Russia to the tundras or Arctic wastes of Asia, chiefly within the Arctic Circle. In winter it is often common in our marshes, coming in October and varying greatly in numbers from year to year. This was one of the species of which the eggs were first found by John Wolley in Lapland, and his description of the habits tallies with those of our common bird; in the shooting season, however, it lies still more closely, rises without any alarm-note, and flies for shorter distances. In colour they are somewhat alike, but the Jack Snipe shews a considerable amount of metallic green and purple above. The eggs are smaller and, on the whole, have darker brown markings.

The Dunlin (*Tringa alpina*) is one of the first shore-birds to be noticed by the novice, for in early autumn it may be seen running about the wet sands in large or small flocks, which are often so tame that he can actually walk among them, though it is not long before they become more wary. These flocks are mainly composed of migrants from abroad, but the bird breeds on the coasts of northern England as well as on elevated moorlands from Cornwall northwards; in Scotland it is much more abundant in summer, while in Ireland it nests in a good many localities. Eastward it is found from Holland and north Germany to Siberia, and thence northward, as well as in North America, Iceland, the Færoes, and rarely in Spain and Italy. Several races have been distinguished, but cannot be considered in our limited space. The food is of insects, worms, crustaceans, and so forth; the note varies from a "tweet" to a whistle; the flight is swift and often circling. The nest, placed in short herbage or stunted heather, is a slight structure, chiefly of grass ; the four eggs are pointed, rather long, and of a buff or greenish colour with red-brown markings. The plumage in summer is more or less chestnut and black above, varied with grey; the breast is black, the belly white, the fore-neck whitish with brown stripes. In winter the black breast is absent, and the upper parts are ashy grey. Plover's Page, Oxbird, Purre, and Sea-snipe are alternative names, while all the small Sandpiper tribe are indiscriminately called "Stints." It will be noticed that many of the rarer species, which breed chiefly or entirely in north Russia and Asia, are, naturally, most common during migration on our eastern coasts.

The Little Stint (*T. minuta*), which breeds in the Arctic regions from north Norway to the Taimyr River in Asia, visits us on passage in autumn and spring, sometimes in considerable numbers. The small flocks chiefly follow our eastern coast-line, and generally linger from August to early October, after which they are not seen again till May. Their habits resemble those of the Dunlin, but the note is almost like that of the Swallow ; the nests are mere depressions in the soil with a slight lining ; the eggs are miniature specimens of those of the Dunlin. The plumage in summer is rufous and black above ; the breast is ruddy with small brown spots, becoming white towards the belly. In winter the bird is ash-brown with entirely white lower parts.

Temminck's Stint (*T. temmincki*) is about the same size as its congener, but is differently coloured, being greyish brown on the upper parts, with dark streaks and blacker bars, while the under parts are white, except the throat and breast, which are brownish with deeper markings. In winter this species resembles a miniature Common Sandpiper (p. 210). It is rather rare in Britain, occurring on passage in autumn and spring, chiefly in the south-east and south of England; but it breeds much nearer our shores than the Little Stint, in Norway, Sweden, Lapland, Russia, and Siberia, the first eggs known having been taken at the head of the Baltic. They are rather longer and paler than those of the last-named, and the nest, placed in marsh growth, is somewhat more substantial. The note is sharp or at times of a trilling nature, while the bird is much given to hovering in the air near its breeding-quarters.

The Curlew-sandpiper (*T. ferruginea*) is not uncommonly met with on the autumn and spring migrations at various points of our coast, generally in company with Dunlins. Above it is chestnut, black, and grey, below almost entirely brick-red. But this summer plumage is seldom present in British-killed specimens, for by autumn the red colouring vanishes and the lower parts become white. The bill is long and decurved, as in the Curlew, the rump white with black bars; in a flock of Dunlins the latter fact and the bird's greater stature render it clearly distinguishable. It was not until 1897 that Mr Popham found the eggs of this species at the mouth of the Yenisei, though breeding individuals had been observed earlier in other parts of Arctic Asia and as far west as Kolguev Island; it is impossible, therefore, at present to delimit its summer range. The eggs are somewhat like those of the Snipe, but much smaller; the nest, as might be expected on the Arctic wastes, much slighter. The habits, as we know them, are those of the Dunlin, though the flight is stronger and the notes more pronounced.

The Purple Sandpiper (*T. maritima*), which is common on many parts of our coasts from September to May, and has even been supposed to have bred on the Farne Islands off Northumberland, prefers rocky shores to sandy, especially where the rocks are covered with sea-weed. There it is to be found in company with Turnstones, Redshanks and Oyster-catchers, while it comes further south in October. The colours above are purplish, black, brown, and rufous; the lower parts are white with greyish breast and much brown spotting. In winter the bird is browner. Abroad it breeds freely in the Færoes and Iceland, as well as on the Scandinavian coasts, and round the whole of the Arctic regions, if we include a race from western America and take note of the fact that the Asiatic tundras (or Arctic moorlands) do not suit its habits. All these smaller Sandpipers being more or less similar in flight, food, and so forth, we need only call attention in the case of the present species to its great tameness in winter, its custom of searching spray-washed rocks for food, and its reiterated double cry. In the north it nests on the hills or at sea-level; the eggs, which are laid on a little grass or some dead leaves, are green or olive with reddish and purplish markings.

The Knot (*T. canutus*) is another of those species of which the eggs have only in recent years been certainly identified, though supposed specimens have been more than once described. In 1901 Dr Walter, in 1902 and 1903 Mr Birulia and other members of the Russian North Polar Expedition brought well-authenticated eggs to Europe from the Taimyr Peninsula and the New Siberian Islands in Arctic Asia respectively, but there can be no doubt of the bird breeding in north Greenland and Arctic America also, for it has been found in summer and its downy young picked up in more than one district. It visits our shores in great numbers between late August and May, being then grey above and white with darker spots below. In its northern home it is black above with chestnut and white markings, and uniform chestnut below, except for the flanks and tail-coverts, which are black and white, and the head and nape, which are streaky reddish brown. The flight is stronger than in most of our smaller waders, but the food in winter is similar; in the far Arctic regions it subsists as best it can on insects and buds of lowly plants. Large flocks are usually formed on our coasts, and the birds are then very quiet; in summer they are said to soar and clap their wings over their backs. Knots used to be fattened for the table in the seventeenth century.

The Sanderling (*Calidris arenaria*) visits us at the same seasons as the Knot, but only a small proportion of the migrants remain during the winter. The flight is seldom long sustained, and the flocks often circle round again to the feeeding ground from which they have been disturbed, while they have a decided liking for running at the edge of the tide on sandy shores. Whether it be summer or winter the food resembles that of the Knot, and both breed in the far north of the New and the Old World, but the Sanderling has also been reported to do so in Iceland. The note is monosyllabic and shrill; the nest is of the slightest description, the eggs are pale greenish with dull brown spots and are curiously like light specimens of those of the Curlew, except in size. This species is pretty common in Britain in its grey winter plumage with nearly pure white under parts; but it is usually coupled with the Turnstone as one of the most cosmopolitan of birds, being found at that season as far south as Australia and Patagonia. In summer it is chestnut and black above and white below, with pale reddish chest spotted with brown. It has no hind-toe.

The Ruff (*Machetes pugnax*) is so called from the extraordinary erectile ruff of feathers round its neck

Ruff

which is lost after June and gradually grows again with the head-tufts in the spring-time. The range of body-coloration is remarkable, including every variety of black, brown, chestnut, grey, and white, with a white belly; but each male bird regains the same colours annually. The female is called a Reeve, and resembles the male after his summer moult; she is brown varied with buff above, and lighter below, with the usual white belly. This species occurs from autumn to spring on our flatter shores, though rarer in the west, and an occasional pair appears still to nest on the Norfolk Broads. Abroad its summer range includes Scandinavia and Russia, Holland, north France, north Germany, Poland, and northern Asia. The male's flight, laboured in the "time of ruff," is at other times swift and direct, as is that of the female; the note is a low reiterated " whit "; the food is both of insects and seeds, with worms and no doubt small creatures of the shore, where the birds are sometimes found in moderate-sized flocks. They used to be caught in huge numbers in the fen-lands for fattening both in autumn and spring; in the latter season the Ruffs were netted when " hilling," that is, gathering on the drier knolls to shew off to the Reeves, and sparring, as polygamous birds are wont to do. The four olive-green eggs with somewhat oblique brown markings are laid in short herbage on a slight bed of grass.

The Common Sandpiper (*Tringoides hypoleucus*) is not a shore-bird, though found near the mouths of tidal streams on migration about September. It comes to us in April, and soon occupies its breeding grounds, which are typically on the stony stretches of fast-running rivers and on lakes in hill-districts. Naturally therefore it is seldom found in southern and south-eastern England, but elsewhere its short sharp flight before an intruder, its plaintive trill, and constant state of excitement near the nest are well-known characteristics. There is nothing unusual in the food of insects, worms, and so forth, but the bird can hover a little, uttering a sort of song. The foreign range extends from the Arctic Circle to the Mediterranean, the Atlantic Islands, the Himalayas and Japan; in winter the bird migrates as far south as Australia and Cape Colony. The nest consists of a few bits of grass or leaves, and is usually placed among shingle under the lee of some plant, fallen log, or stone, but not uncommonly in a field or on a bank; stony sides of lakes or rivers, or islands in the latter, are the favourite sites. The four pointed eggs are exceptionally broad at the larger end and are of a stone-colour with purplish or reddish brown and grey markings. In summer the bird's upper parts are greenish brown with darker marks and a good deal of white on the conspicuously rounded tail, the breast is more ash-coloured, and the belly white; in winter the upper parts are purer brown.

We now come to a group of larger Sandpipers, all of which may be included in the single genus *Totanus*. The first two are the Wood Sandpiper (*T. glareola*) and the Green Sandpiper (*T. ochropus*), both brown birds with buff-white spotting and white lower parts, the fore-neck and breast being profusely streaked with brown and the upper tail-coverts nearly white. Though the adult of the latter species is slightly greener above, identification would be difficult were it not for the pattern of the axillaries and central tail-feathers, well illustrated in Saunders' *Manual of British Birds* (ed. 2, p. 612). The axillaries, which spring below the wing, in the Wood Sandpiper are almost white with brown flecks, in the Green Sandpiper blackish with narrow white bars meeting to form "arrow-heads," and the tail-feathers in the first-named have very much narrower black bars than in its congener. It is a curious fact that, though the Green Sandpiper is much the commoner bird with us inland on its passage in autumn, and sometimes stays for the winter, yet on the spring passage it has never remained to breed, while the Wood Sandpiper's eggs were once taken by Hancock at Prestwick Car, near Newcastle-on-Tyne. No other instance of the nesting in Britain of either species can be substantiated, though such have been reported. The Moray Firth seems to be the northern limit of their range in our country, as is the Arctic Circle in Europe and Asia; but that of the Green Sandpiper extends to Holstein, Germany, central Russia and central Asia, while that of the Wood Sandpiper includes Holland, the Danube valley and even parts of southern Europe. The habits and food are those of the last genus, but the note is shriller and more insistent in summer, and both the birds, at least from Germany eastward, have been proved to use also old nests of Thrushes and the like in which to deposit their four large greenish or buff eggs with red-brown spots and blotches. The same fact has lately been proved for their American relation the Solitary Sandpiper, but the habit is much less constant in the Wood Sandpiper.

The Redshank (*T. totanus*) breeds commonly both inland and near the coast, where it can find marshy grass-land suitable to its requirements; some thirty years ago most of its colonies were by the sea, often on salt-marshes, but of late it has spread to districts far inland, while it is not uncommon even in Shetland. Undoubtedly many migrants arrive in autumn from abroad and join the home-bred birds, which at that time of year seek the shore; the majority then depart for the south, but a certain proportion remain during the winter. The small flocks no doubt vary their usual insect-food by feeding on small marine creatures, for their regular haunts are the rocks covered with sea-weed, though at high tide they pass up the rivers and streams to the neighbouring meadows or saltings. The foreign range includes all Europe, the extreme north of Africa, and also northern Asia, though in the last-named it does not extend northward to about the Arctic Circle as it does in Europe.

Redshank's nest

The Redshank is a most vociferous bird, and, when itself disturbed, disturbs all its kin; in the breeding season it is specially noisy, hastening to meet an intruder and seeing him "off the premises." The nest of a little herbage is cunningly hidden in a tuft of grass or rushes, which are often twisted together to hide the four stone-coloured eggs marked with purple-brown; more rarely the contents are open to view.

The flight is, or can be, particularly swift; at the breeding quarters the parents fly backwards and forwards or circle around in a state of the greatest excitement. The bill, except at the tip, and the legs, are orange red; the general plumage is brownish with darker streaks and bars; the rump is nearly white, and there is so much white on the tail and lower parts that it shews more than the darker colours do when the bird is on the wing.

The Spotted or Dusky Redshank (*T. fuscus*) is an irregular visitor to our seaside or inland marshes on both migrations; it has even occurred in Orkney, but does not remain for the winter. Being rare in Britain, the question of its nesting-quarters used to be of paramount interest, until Wolley discovered them in Finland in 1854. Since then the bird has been proved to breed in other parts of Europe north of the Baltic and across northern Asia to the neighbourhood of Kamtschatka. It is black above and below with small white spots, the rump and tail alone shewing much of that colour. In summer, therefore, it is a far darker bird than the Common Redshank; even in winter the grey of the plumage is darker in parts, while the legs are always livid, with crimson round the joints. The eggs are more beautiful and often greener than those of its congener and are laid in somewhat drier situations in the vicinity of a marsh, while the hen at times sits very closely. It is also a less vociferous bird, and in habits comes into comparison with the Greenshank rather than the Redshank, though it is more fond of perching on trees than either.

The Greenshank (*T. nebularius*), which breeds in the Hebrides and on the mainland of Scotland as far south as Perthshire and Argyllshire, if not occasionally in the Border country, does so abroad in Scandinavia, north Russia, and Siberia. The olive-green legs distinguish it at once from the Redshank, while it is a larger bird and decidedly greyer above. The general habits of the two species are identical, but the Greenshank is quieter and has a note sufficiently different to be appreciable by those accustomed to meet with both waders. Moreover it does not frequent sea-weed covered rocks, but muddy or sandy shores and salt-marsh creeks. Many individuals are met with inland on passage, though comparatively few remain to breed; the slight nest of herbage or dead leaves is built on the side of some forest-pool or on comparatively dry spots near a loch or other piece of water. The four eggs are larger and lighter than those of the Redshank with more purplish markings.

The Bar-tailed Godwit (*Limosa lapponica*) arrives on our coasts in August, between which month and the following June larger or smaller flocks frequent our sandy or muddy shores and estuaries. The majority pass southward for the winter, but many individuals remain with us; while in different localities they differ much in abundance. The upper parts are brown, with black markings on the back and much white on the rump and tail, the rest of the plumage being of a fine brownish red with blackish streaks on the head; towards winter this fine ruddy colour gives way to white below, and the whole of the upper parts are brown and grey, except the tail-feathers, which are always barred and give its name to the bird. This is our common Godwit, though it does not breed in Britain, but only from Lapland and northern Russia to the Taimyr Peninsula in Asia, east of which a slightly different form takes its place. The flight, food, note, and eggs are similar to those of the next species, but in the Arctic regions the localities chosen for the nest, when any is made, are naturally of a more barren description.

The Black-tailed Godwit (*L. limosa*) is now but an uncommon visitor to any part of Britain between autumn and spring, though it bred in Yorkshire, Lincolnshire, Cambridgeshire, and Norfolk during the first half of last century. It is one of those species which might possibly nest with us again, in the absence of persecution, for it does so as near our shores as Holland, while its range includes the Færoes, Iceland, northern Europe, and northern Asia, if we include a slightly different form found in the east of the latter. It differs from the Bar-tailed Godwit in having the head, neck, and breast reddish buff with a certain amount of black markings, which are more evident on the brown back, but the chief distinction lies in the tail, which has only one wide black bar. The rump and a wing-bar are white, the belly white and brown. In winter it is chiefly red-brown above and grey below. The flight is fairly strong, and the birds travel well when disturbed on the sands or mud flats which are their resorts on passage; the cry is of two syllables and of a yelping nature. The food resembles that of Sandpipers generally; the nest is a slight fabric placed in rough herbage; the four pear-shaped eggs are of a peculiar dull greenish brown colour with subdued brown spots.

The Curlew, or Whaup of Scotland and its borders (*Numenius arquata*), is a most characteristic moorland bird, occurring on our hills and low-lying heaths from south-western England and Wales to Derbyshire and the north country generally. It is less common in Orkney and Shetland, hardly occurs in the Færoes or Iceland, but is pretty abundant in northern and central Europe and thence reaches western Asia, where it meets a very similar ally. The Curlew is a streaky brown bird with a white rump and an extremely long decurved bill; it has a prolonged melancholy cry,

Curlew on nest

easily recognised but hard to describe; it feeds on insects, worms, and so forth, as well as on berries from the moors and marine animals from the shore; the nest is hardly more than a large shallow depression in the heather or grass, and the four fine pear-shaped eggs are olive-coloured or greenish with bold brown blotches or smaller spots.

When breeding the parent birds circle anxiously round an intruder, and come very near him after the young are hatched, but from autumn onwards they are the most wary of shore-birds and practically act as sentinels of the beach. It is uncertain how many reach us from abroad, but non-breeding birds may be found by the sea at any time of year. The name, as pronounced in the north, "curr-loo," is no doubt derived from the note, while the weird cry and curious appearance of the bird caused it of old to be regarded with superstition by the Scots, who averred that it was a wizard, and could only be killed with a silver slug.

The Whimbrel (*N. phæopus*), which resembles a small Curlew, but has a dark brown head with a wide pale central stripe, is a somewhat more Arctic species than its congener, being found in summer in the Orkneys, Shetlands, Færoes, Iceland, the fells of Scandinavia, and over the districts north of forest-growth both in Europe and western Asia, the eastern Asiatic form differing in its more streaky rump. Many migrants arrive on our coasts about May, and a few stragglers may remain, but towards the end of summer all return from their northern haunts; they are easily recognised by their constantly repeated shrill whistling notes, dimly recalling those of the Curlew. The food of the two species is alike, but the flight of the Whimbrel is less prolonged, while it has a curious habit of executing spiral movements near the nest. On the whole the eggs are brighter green or olive and their markings more distinct; the parent birds, moreover, are decidedly less shy.

ORDER XIV. GAVIÆ

This Order is here taken to contain the Families *Laridæ*, with the Gulls and Terns, and *Stercorariidæ* or Skuas. They are all seafaring forms, though some breed inland instead of on our shores. The bill is strong, horny, and curved in Gulls; still stronger and hooked in Skuas, where it has a hard "cere" or membrane at the base; long, straight, and pointed in Terns. The feet vary in length, the front toes are webbed, the hind-toe is elevated. The wings are long, particularly in Terns; the tail is moderate and even in most Gulls, but in Sabine's Gull is forked, as it is in Terns, where it is prolonged into streamers; Skuas have the two central feathers elongated. The downy young vary in colour from white and grey to yellow and brown, and are usually mottled or striped; they lie sluggishly in the nest for some time, though they often leave it and wander about in the day.

Family LARIDÆ, or Gulls and Terns

The Gulls and Terns together compose one Family, though a Subfamily *Larinæ* is often used for the former, and *Sterninæ* for the latter. But, while combined for anatomical reasons, their habits differ to a considerable extent. Gulls have a steady and often gliding flight and for the most part keep high in the air, if they are not resting on the ground or on the surface of the water; Terns, often called Sea-swallows, have a much more energetic way of moving, and give an impression of extreme buoyancy as they hover, dart, twist, and turn, while uttering their shrill or more or less grating cries. All is excitement, as opposed to the ordinarily placid conduct of the Gull. It is not their habit to repose on the water, but they skim along over the tops of the waves, darting down to secure a fish and not uncommonly submerging themselves by the violence of their sudden plunge. Sand-eels and the like from shores left dry by the tide form part of their diet, but those Terns and Gulls that breed inland have a much more varied bill of fare. Gulls, indeed, are practically omnivorous, for, in addition to the creatures of the sea, they feed on worms, grubs, grain, and even berries, not to mention the young birds and lambs destroyed by such bold robbers as the Greater Black-backed Gull. It is a pretty sight to see flocks of Gulls—usually of the Black-headed or Common species—following the plough; they keep close behind the ploughman, rising in a cloud from time to time for no obvious reason and settling down again immediately to resume their interrupted repast. In many towns, Perth being a striking instance, they sit on the roof-ridges in the early morning waiting for any refuse that may be thrown out; later in the day they are equally on the alert as scavengers of the quiet streets. In Perth the Black-headed Gull, which breeds close at hand, is the bird in question, but in other towns near the sea the Common Gulls, or less often the larger species, crowd upon the tops of the suburban "dust-heaps" and fight with each other for the delicacies they find there. Gulls do not scream like Terns, but utter querulous or mewing cries, which in the case of the largest species become harsh croaking sounds. Both are distinctly gregarious on the coast, as well as at their nesting-colonies, and there is little difference in this respect; perhaps Terns congregate more closely when sitting on the shore, Gulls when on the cliffs or inland marshes. But all the inland species have habits that vary somewhat from the normal, and these will be considered under the several headings. Gulls build a large nest of grass or sea-weed, mixed with refuse, or, on lakes, of water-weeds, rushes, and flags ; they lay two or three eggs of a greenish or olive-brown colour with spots and blotches of dark brown and grey. Terns scrape a hole in the turf, among shingle or sand, or even make use of rocks, and add little or no lining, the two or three eggs being cream-coloured, green, or brown in ground-colour with bold markings of brown, grey, or black. The Black Tern and its congeners are exceptional in making nests in marshy places of water-weeds, which sometimes almost float on the surface; the Sooty and the Noddy Tern (p. 261) in laying a creamy white egg with lilac and brownish red spots. The latter commonly makes a nest of grass and rubbish and places it on the top of a low tree or bush. In conclusion it should be observed that, except for stray individuals, no Terns remain with us in winter.

SUBFAMILY **Sterninæ,** OR TERNS

The Black Tern (*Hydrochelidon nigra*), which is blackish grey rather than black with red-brown feet and black bill, and is somewhat lighter above, used to breed until the middle of last century in Norfolk, and at earlier dates in Lincolnshire and Cambridgeshire, besides several parts of East Anglia, where it was called the "Blue Darr" or "Carr-Swallow." It is now only a visitor, which occurs at various times of year and sometimes remains as late as May. Abroad it ranges from Holland, Denmark, and the Baltic to the Obb River, Lake Zaisan and north Africa. The food consists largely of insects, but is varied by fish, newts, frogs, and leeches; the nest of rotting weeds is always in a marsh, and the eggs are usually dark coloured.

The White-winged Black Tern (*H. leucoptera*) is a blacker bird with red bill and feet, white tail, and white wings that become grey from about the middle to the tips; it is a rarer visitor to Britain than the Black Tern, but is seen at similar seasons and has similar habits. From Poland and the Rhone valley it appears to extend to north Africa, while it ranges from the Caspian Sea to China, but further information seems necessary before we can lay down certain limits.

The Whiskered Tern (*H. leucopareia*) is quite irregular in its visits to our shores; it is as much a marsh Tern as its congeners, but in colour is grey with a black head and belly, red bill and feet and a white cheek-stripe. The range is more southerly, comprising the south of Spain, southern Russia, Turkey, Greece, and north Africa, and extending thence to north India and perhaps Mongolia. A few pairs breed in the south of Germany, Poland, and at the mouth of the Rhone.

This species is given here merely for purposes of comparison, while the Gull-billed Tern and the large Caspian Tern must certainly be left to the subsequent list of uncommon migrants; but we may note that both have bred as near to Britain as the island of Sylt off the shores of Denmark, and possibly may be found more frequent visitors to our shores than has been supposed. They belong to the following genus, *Sterna*, and not to *Hydrochelidon*.

The Sandwich Tern (*Sterna sandvicensis*), the largest of the species which breed with us, does so regularly on the Farne Islands off Northumberland, Ravenglass in Cumberland, and thence northwards in certain localities as far as the northernmost of the Orkney Islands. Besides this, it has been known to nest in more southern places and in Ireland, and it is hoped that many of the colonies will be permanent, though the bird is apt to change its quarters for no very apparent reason. Abroad it ranges from Holland and Denmark to the Caspian Sea, as well as from north Carolina to Mexico and Honduras in the western hemisphere, though the American form is slightly different. It is also known to breed in Spain, some of the Mediterranean Islands, Tunisia, and the eastern Canaries. The colour is grey above and pinkish white below, the crown, nape, legs, and bill being black, with a yellow tip to the bill. Two eggs of a whitish or buff colour, with black, grey, or red-brown spots and blotches, are usually laid on a sandy shore, but sometimes among short grass or sea-plants on an island.

The Roseate Tern (*S. dougalli*) still breeds off the shores of Anglesea, in Ireland, on the Farne Islands, and in the Moray Firth, while a few pairs may do so even in East Anglia, the Scilly Islands, and elsewhere, though they are supposed to have left those parts. It is found in summer from western France to the Mediterranean, from New England to Venezuela, in the Azores and Madeira, and down the Indian Ocean, while it also nests in Madagascar, Ceylon, the Andaman Islands, and so to south China and Australia. But for the exact localities the reader must consult the pages of one of the larger works on birds. Grey above with a black head, as in the case of its congeners, the lower parts in life are of a beautiful pink colour, which fades considerably after death; the bill is black, the feet red. The two or three eggs, laid on sand or rock, resemble those of the Common Tern, but are generally buff in ground-colour; the note is comparatively harsh and grating.

The Common Tern (*S. hirundo*) is the most abundant species in England and southern Scotland generally, but further north the Arctic Tern (*S. paradisea*) out-rivals it in numbers, and was till lately the only member of the genus known to breed in Shetland. Both occur up the west coast and in Ireland. If we

Common Tern

omit exact details, the Common Tern ranges over temperate Europe and western Asia, with the Atlantic Islands, north Africa and North America, while only the Arctic Tern is found in the Færoes, Iceland, and the more northern regions. In the Common Tern the head and nape are black, the upper parts grey, the lower greyish white; the bill is orange with yellowish white tip, the legs are red. The nest and eggs are of the usual type, described in the general account above.

Gaviæ

The Arctic Tern has a circumpolar range, though in eastern Asia a closely allied form is more abundant; it has an entirely red bill, and the outer feathers of the deeply forked tail are longer than in the common species. On the whole the eggs are smaller, more often green in ground-colour, and occasionally fine yellowish red.

The Little Tern (*S. minuta*) is much more local than either of the two birds just mentioned, and almost always lays its smaller eggs among shingle or in sand, generally nearer the sea than its congeners. They are light buff or greenish white with grey and brown markings, and do not vary much; the note is shrill and the parents are bold and restless. This Tern breeds in various localities from Orkney to the extreme south of England, and abroad from the Baltic to north and west Africa and north India. The limits cannot, however, be yet determined with certainty, as more than one similar species meet our bird in Asia, and their ranges not improbably overlap. The smaller size, white forehead, orange feet, and yellowish orange bill with black tip distinguish it at once from its British congeners.

SUBFAMILY **Larinæ,** OR GULLS

The first Gull to be mentioned is Sabine's Gull (*Xema sabinii*), our only fork-tailed species; it is comparatively small, has a lead-coloured head and neck, with first a black and then a white collar, white tail, and undersurface. Many individuals have been seen or killed at various points of our coasts in the latter half of the year, and the bird is known to breed in America, from Greenland to Alaska and along the Arctic coasts of Europe and Asia from Spitsbergen to east Siberia. The appearance of individuals in summer in the Polar regions of Europe make it almost a certainty that this Gull ranges all round the Polar seas. It can hardly be said to have any nest, for the two olive-brown eggs with dull brown blotches are reported to be laid either on the bare ground or on tussocks in marshes.

The Little Gull (*Larus minutus*) is another of our uncommon visitors, but instead of occurring only at considerable intervals, it seldom fails to put in an appearance in the autumn or winter. At times single individuals only are obtained, but in some seasons large numbers are present on our shores and may be found even as far north as Shetland. The bird breeds inland from Jutland to the south-eastern Baltic districts, northern Russia, and temperate Asia, and makes a nest of grass and so forth on the drier parts of large marshes, the three or four eggs being miniature examples of those of the species next to be described, though usually of the browner type.

The Black-headed Gull (*L. ridibundus*) always breeds inland in Britain, and its eggs are often gathered for eating; they are usually olive-brown in colour with blackish or brownish markings, but vary—though less than those of many other Gulls—to green or greenish blue. A red Gull's egg is exceptional in any case, and therefore much prized by collectors. The nest of marsh herbage is of considerable size and is built on flat islands or in almost inaccessible parts of swamps, more rarely among "casual" water in badly drained spots. Whereas the Little Gull has a decidedly black head and upper neck, the present species has really only a brown hood, not reaching down the neck behind, while it may be noted that two uncommon stragglers in our subsequent list, the Mediterranean and Great

Gaviæ 227

Black-headed Gulls, have perfectly black heads, as in the Little Gull. Sabine's and the rare Bonaparte's Gull are our only other black-headed species, and in all the head becomes nearly white after the breeding season. Both the Little and the Black-headed Gull are grey above and pinkish white below, but the smaller bird lacks the black on the outer wing-quills, though it is black below the wings. In both the legs are red, of a more purple hue in the Little Gull. Colonies of the Black-headed Gull are comparatively rare in the south and south-west of England, commoner in Ireland and Scotland; but they are even found in Shetland, as well as in the Færoes, temperate Europe, and temperate Asia generally. The note is shrill and peevish.

Of the Gulls without black heads all but the Kittiwake occasionally breed inland, either round moorland waters or on islands in them, and the species which most frequently does so is the Common Gull (*L. canus*), so called from its abundance on our coasts and fields during the colder seasons of the year. No breeding colonies exist south of the Scottish border and few in Ireland, but from western and southern Scotland they increase to the northward, though rare in the eastern half. The grassy nests are usually placed on islands in small lakes, on their stony sides, or on rock-strewn slopes near the sea, and are not found on precipitous cliff-faces; the eggs are of the usual Gull type, but comparatively seldom green. The Common Gull is abundant throughout Scandinavia, the Baltic, northern and central Russia, while the form which extends across northern Asia to Kamtschatka is merely a little larger and darker above. This species is grey, with white head, tail, and under surface; the wing-quills have black ends and the legs are yellow. The note is more mewing than that of most Gulls, whence it is called the Sea-mew.

The Herring-Gull (*L. argentatus*) is really the common Gull of Britain, and is often so called by country-folk; its general coloration is that of the last species, but it is much larger and has flesh-coloured legs. Apart from a few colonies on moorlands and Scottish lochs, it is found breeding on cliffs or rocky islands in all

Herring-Gull

suitable parts of Britain. It ranges generally from Arctic Europe to north France, as well as across Arctic America from Greenland to central Alaska. In winter it migrates far to the south. In the Atlantic Islands, the southern European coasts, and thence to the Black Sea and Lake Baikal the Yellow-legged Herring-Gull takes its place. The note is shrill and querulous but not so like a "mew" as that of the Common Gull, while the eggs are similar, but larger.

The Lesser Black-backed Gull (*L. fuscus*) is found in summer from Devonshire and Cornwall up the Welsh coasts to the Isle of Man and the west of Scotland generally. It also breeds inland in Northumberland and Cumberland, as well as in Ireland, but such cases are exceptional. On the east coast of England it is the common Gull of the Farne Islands, and ranges thence in smaller numbers to the north of Scotland. The nest and eggs are indistinguishable from those of the Herring-Gull, but are most commonly on flat islands or sloping ground, and not on cliffs, for which reason the colonies are somewhat local. A few exist even off Morocco, in the Mediterranean and on the west of France, but the chief summer haunts are northward, and extend from the Færoes and Scandinavia through the Baltic to Russia. The British form has just been subspecifically separated from the type. The mantle is black or very dark grey; there is a little white on the wing-quills; the bill is yellow with some red at the gape, as in the Herring-Gull; the legs are yellow; otherwise the bird is white.

The Great Black-backed Gull (*L. marinus*) is much rarer than the preceding species, and is as often found in single pairs as in colonies; a few birds breed in the south-west of England and in Wales, but by far the greater part do so in the west and north of Scotland. Mainland cliffs are not uncommonly utilized, but on the whole rocky stacks and outlying islands are preferred, and in Scotland, the Lake District, and the Solway Firth even marshes, often at considerable elevations. In Ireland the bird is not rare. Abroad it ranges from Baffin Bay and Labrador to Greenland, the Færoes, Iceland, and Europe north of the Baltic, as well as to the Yenisei river, while it is known to breed in north-western France. The two or three eggs are nearly always stone-coloured, but are often very boldly marked with dark brown and grey. The coloration, apart from a few differences in the white markings on the wings, is similar to that in the Lesser Black-backed Gull; but the legs are flesh-coloured, and the size of the bird is much greater.

The Glaucous Gull (*L. glaucus*) and the Iceland Gull (*L. leucopterus*) do not breed in Britain, but visit us regularly in considerable numbers towards winter, especially in seasons of extreme cold. They are both very white-looking birds with pale grey mantle and wings, but the former is much larger than the latter, and in this respect they correspond with the two species of Black-backed Gulls. The Glaucous Gull nests through the Arctic seas of both worlds and in Iceland; the Iceland Gull does not breed in that island, but in Jan Mayen, Greenland, east Arctic America and no doubt other parts. Perhaps the best distinction, besides the size, lies in the pink colour of the legs in the first-named, as opposed to a flesh-coloured, or yellowish tint, in its ally. The eggs are much as in the Greater Black-backed Gull, while the nest is commonly, but not always, on precipitous cliffs; the food is of the most diverse description, and includes berries of northern shrubs.

The Kittiwake (*Rissa tridactyla*), our most dainty and delicate-looking species, can be distinguished at once from its British congeners by its black legs and by the (apparent) absence of a hind-toe. The mantle is of the same deeper grey as that of the Common Gull, which the bird most nearly resembles. It always breeds in large colonies on sheer precipices, the nests of turf, sea-weed, straw, and like substances being fitted into small angles in the face of the rock, where at first sight

Gaviæ 231

Guillemots and Kittiwakes

there appears to be little or no room. A picture of Kittiwakes on their nests, therefore, shews the sitting birds almost clear of the cliff. The eggs are greyish-white, stone-coloured, or light-brown, richly marked with brown and grey of various shades; they are much prettier than those of ordinary Gulls, and do not so often tend to a green ground-colour. This species is found locally all round our coasts, and is rarely driven inland by storms; abroad it breeds in Brittany (with the Channel Islands), and thence to the Færoes, Iceland, and all the circumpolar regions, if we do not separate a very doubtfully valid form in Bering Sea, and allow for a really distinct species in the north Pacific Ocean. In some respects the Kittiwake recalls a Tern, for it dives and swims under water in a way that few Gulls attempt, while the flight is more wavering, and it has a distinct habit of hovering. A black band near the end of a Gull's tail is a sign of immaturity.

Family STERCORARIIDÆ, or Skuas

Of late years a great deal has been heard of the Great Skua (*Catharacta skua*), on account of the necessity for its preservation in Shetland, where alone in Britain it breeds. It is a large and heavy dark brown bird with a tendency to yellow on the neck and a little white on the back and wing; it lays its two dull olive eggs with darker spots and blotches in hollows which it scoops out in the coarse grass and short heather on the tops of the hills. If a visitor approaches the colony the parents swoop down upon him from the front, apparently aiming at his face; just before reaching him, however, they drop their feet and rise above his head. Probably he may have raised his arm or a stick, which helps to produce this effect, but the action of the birds seems to be natural, and it is doubtful if they really mean to strike. But they look very terrible, and generally manage to frighten new-comers. When not nesting the Great Skua is a seafaring species, which keeps well out in the ocean and is comparatively seldom observed with us except in northern Scotland ; it lives, as others of this group do, on fishes and fowl, and constantly procures the former by forcing the Gulls to disgorge what they have just caught. Its flight, though heavy, can be easily accelerated with this object ; its note is a harsh cry, supposed to resemble the syllables " skua, skua." The present species is the biggest of our "Parasitic Gulls" or "Gull-teasers," and is termed the Bonxie in Shetland; elsewhere it only breeds in the Færoes, Iceland, and in Hudson Strait, though closely allied species are found in the southern ocean down to the Antarctic regions.

Our remaining Skuas belong to a different genus, of which Richardson's or the Arctic Skua (*Stercorarius parasiticus*) is the only one that breeds in Britain, being there confined to the extreme north of Scotland with its islands, though sufficiently common in Shetland. The Great Skua group have the central tail-feathers only slightly projecting, but all the forms of the second group have them extraordinarily lengthened. The Arctic Skua has two phases of coloration ; in one the bird is entirely sooty-brown, in the other it has the lower surface and neck white ; in both a yellowish tinge shews on the cheeks and neck. These phases cannot be called sub-species or local races, for they interbreed with each other, though either may be the commoner at any one locality. Colonies are formed which nest on grassy or mossy moorlands, not necessarily on the tops of hills; the bird's habits are similar to those of the Great Skua, except that they usually attack an intruder from behind and sometimes actually slap him with their wings, while they customarily try to entice him from their eggs or young by grovelling on the ground before him, and uttering most distressful mewing cries. Their

Arctic Skua

lesser size causes them to chase the smaller rather than the larger Gulls for fishes, and especially the Kittiwake. Their eggs are of a brighter brownish or greenish colour than those of the Great Skuas. After the breeding season the Arctic Skua migrates down our coasts, and even reaches the southern oceans; in summer it is found in the Færoes, Iceland, and thence all round the Pole in the more Arctic regions.

Gaviæ 235

The Pomatorhine Skua (*S. pomatorhinus*) is merely a visitor to our shores from autumn to spring, while it is much more abundant in some years than others, and generally moves further south for the winter. It undoubtedly breeds on the Arctic coasts of Asia, where eggs have been taken in more than one locality, and has been reported to nest in northern Europe, as well as in Greenland and Arctic America. For habits the reader may consult those of the last species. As regards coloration, dark examples are only found exceptionally, while the birds have white necks and breasts, with a yellow tinge on the former, the crown being black and the central tail-feathers curiously twisted to face sideways.

Buffon's or the Long-tailed Skua (*S. longicaudus*) comes to us in much the same fashion as its preceding congener, but it breeds as near Britain as the fell-tops of Norway and Sweden, as well as all round the Polar shores and islands of both worlds. As it is the smallest of our Skuas, the eggs and nest are naturally smaller also, while the former usually have a greenish tinge; the flight is light and elegant, and the note is said to be more shrill than in Richardson's Skua. The food in the far north may include berries to a small extent, but this habit is not unknown among Skuas generally; still this species seems to be less particular than its larger relatives, as it will eat insects, crustaceans, and worms. The blacker head, greyer back, longer tail-feathers, and smaller size distinguish it from the Arctic Skua, whereas the Pomatorhine Skua is not only larger, but has its remarkable twisted tail-feathers to guarantee it.

ORDER XV. ALCÆ

In the Auks the body is heavy and compact with a rather large head, and the plumage is close and elastic. The stout bill varies considerably, as will be seen below, and is often compressed; the foot is short, with long and webbed front toes and no hind-toe; the wings are absurdly short, the Great Auk having been quite flightless; the tail is also very short, and the birds commonly sit upright on their feet. The nestlings are covered with down, which may be black, grey, or brown.

Family ALCIDÆ, or Auks, Guillemots and Puffins

The members of this Family, in which the sexes are alike, are the counterparts, in Arctic and temperate regions, of the Antarctic Penguins, which differ from them completely in structure. The latter, moreover, are flightless, while all the Auks, except the Garefowl, are able to fly. They swim and dive to perfection, and live on fishes and crustaceans. Among British species the Puffin alone occasionally makes a nest, while only the Black Guillemot lays two eggs, the others laying one. The note is a croaking or grunting sound.

SUBFAMILY **Alcinæ**, OR AUKS

The most striking member of the Family is that quaint bird the Great Auk or Garefowl (*Alca impennis*), extinct since 1844, when the last two specimens were obtained at the island of Eldey off the south-west of Iceland. It bred plentifully on Newfoundland and Funk Island in America before 1819, while the last of its certainly known visits to Britain were in 1813, 1821–2, and 1834, in Orkney, St Kilda, and Waterford

Great Auk

Harbour (Ireland) respectively. Of old the bird also inhabited Labrador, southern Greenland, the coasts of Scandinavia, Denmark, and the Færoes, but never extended north of the Arctic Circle. Eighty skins and seventy-three eggs are in existence; the latter are white with black or reddish brown spots or scrawls, and measure no less than $4\frac{3}{4}$ inches in length. The Great Auk was black above and white below, with a white patch in front of the eye, and a much compressed and corrugated black bill. Its remains have been found in ancient refuse-heaps in northern and western Scotland, Durham, and Ireland, besides abroad.

The Razorbill (*Alca torda*) resembles a small Great Auk, without the white eye-patch, but with a white line across the much compressed and grooved bill; it is found from Jan Mayen in the Arctic regions to the Bay of Fundy and Brittany in the Atlantic, but not in the Pacific. It breeds in hollows of cliffs or on ledges, and lays a white or brownish egg, richly marked with black or brown, in company with the Guillemots, which usually outnumber it, especially in their well-known haunts at the Farne Islands on the Northumberland coast. They arrive at these haunts early in April and leave punctually in August, but may be observed in the tideway at all seasons, and are often found dead on the shore in quantities after a severe winter storm.

The Common Guillemot (*Uria troile*), which breeds from Bear Island of the Spitsbergen group to the Baltic and Portugal, as well as on the Atlantic coast of America down to about the same latitude, also occurs as a slightly different form in the Pacific. It is the "Willock" or "Murre" of fishermen and the "Loom" of the Arctic regions, where the "loomeries" of this and the next species are even vaster than in our northern islands. The Farne Islands, where the eggs literally cover the flat tops of the "Pinnacle" Stacks, and Flamborough Head in Yorkshire, are especially well-known localities. The beautiful large pear-shaped eggs are green, blue, or white, with various spottings or scrawlings of black or red-brown. Ordinarily the birds

Guillemots on the "Pinnacles"

lay them on ledges almost too narrow for safety, while the successive rows of sitting Guillemots are a feature of many a precipitous northern cliff. In general habits they resemble the Razorbill, but in colour they are browner with a non-compressed bill. The egg never has a green lining-membrane, as in that species. Neither bird makes the slightest attempt at a nest.

Brünnich's Guillemot (*U. lomvia*), an Arctic bird, which rarely visits us in winter, and has a stouter bill with a white edge to the upper mandible, forms the greater part of the vast loomeries of the far north. The habits are identical with those of the common species and the eggs are indistinguishable. The so-called "Ringed Guillemot" is a variety of the Common Guillemot with a white ring round the eye.

Black Guillemots

The Black Guillemot or Tystie (*U. grylle*) is a beautiful little black bird, with a large white wing-patch and red feet; in autumn it is barred with whitish above and is white below, as are the young, but not the adults, in winter. Common in northern Scotland, it occurs in small numbers in the Isle of Man and Ireland. Abroad it extends from the White Sea to Iceland, the

Færoes, and the Baltic, and in the western Atlantic to Massachusetts. Other forms, nearly or entirely black, occupy the rest of the northern regions. It is a quiet bird, easily overlooked by the non-observant, which is generally seen swimming near rocky islands or cliffs with boulders at their base, and flies but little. It lays two greenish white eggs streaked with brown and grey in crevices between the boulders, or occasionally in holes low down in the cliffs.

The Little Auk (*Alle alle*) is well known in Britain as an irregular winter visitor, often arriving in great numbers on our northern coasts or found dead on the shore in severe weather. It is seldom altogether absent in winter, but is less common in the south, while it is frequently driven inland by storms, which it seems ill adapted to resist. From Grimsey Island off Iceland it ranges in the breeding season to Arctic America, Greenland and the Kara Sea, and lays one bluish white egg, generally faintly marked with rust-colour, in crevices in cliffs or under stones. The name "Little Auk" is somewhat misleading, and suggests a comparison with the very different Great Auk. Least Auk would be preferable, as there are many more than two species in the whole Family. The colour is black above and white below except the neck.

SUBFAMILY **Fraterculinæ**, OR PUFFINS

By name at least everyone knows the Puffin, Sea Parrot, or Tammie Norie (*Fratercula arctica*), a black bird with white cheeks and under parts, orange feet and orange and blue bill. The sheath of this huge compressed bill is shed in pieces in autumn. The bird breeds in suitable places all round our coasts, choosing to burrow in earthy slopes on islands or cliffs, and laying a coarse-grained whitish egg with faint lilac markings on the bare soil or a little dry grass. It is most amusing to walk about a big colony and see the birds popping out of their holes one after the other and

Puffins

speeding like arrows to the sea, which soon becomes covered with them, while it is equally interesting to watch them passing to and fro with fish for their young later in the season. They bite severely if disturbed in their burrows. The Puffin ranges from the Arctic seas in Europe to Portugal and to the Bay of Fundy in

America. It returns very regularly to its breeding haunts in April and is commonly met with in winter at sea.

ORDER XVI. PYGOPODES

This Order contains the Families *Colymbidæ*, or Divers, and *Podicipedidæ*, or Grebes. They are waterbirds of somewhat varying habits, with very close plumage that water cannot penetrate; this is particularly glossy below in Grebes. The bill is very strong and straight in Divers, moderate and sometimes a little curved in Grebes; the feet are peculiarly flattened and have webbed or lobed toes in the respective Families; they are also set very far back. The claws are broad and flat. The wings are decidedly short; the tail is very short in Divers and represented by a few downy feathers in Grebes. The young are downy and of a sooty or streaky appearance.

Family COLYMBIDÆ, or Divers

The Great Northern Diver (*Colymbus immer*) has never bred in Britain, for the supposed instances of its doing so in Shetland have been due to errors of identification. Yet individuals may occasionally be seen in Scotland as late as June, and are frequently abundant on other parts of our coasts. This fine bird breeds in Iceland, Greenland, and North America, mainly south of the Arctic Circle, at least as far westward as the Great Slave Lake; it has a purplish black head and neck with two white bands on the greener throat, which are striped vertically with black; the upper parts are black, regularly spotted with white, the lower parts are white, the irides are crimson, the bill is black.

Very closely allied is the more Arctic *C. adamsi*, a larger bird with a greener tint on the head, a purplish hue on the throat, a yellowish white bill, and more elongated spots. It may breed all round the Pole, and certainly does so in Novaya Zemlya, Arctic Siberia, and North America west of the Great Slave Lake. Specimens have occasionally been obtained in Britain. The habits of both species are similar. The cry is a weird sort of laugh or croak, uttered on the wing as the bird describes wide circles round its breeding-quarters; its flight resembles that of a large Goose, while it dives under water at once if it is alarmed when swimming. Divers, however, are not specially suspicious, and, particularly when near the nest, allow persons to approach within short distances. The fabric, built on an island or at the edge of a lake, is of water-weeds, and may either be a considerable pile or practically non-existent; the two big eggs are brownish olive with blackish spots. In leaving and regaining the nest the sitting bird usually slides along rather than walks, making a distinct track to the water. The food is of fishes, with crabs and so forth from the sea, for Divers only frequent the land in the breeding-season.

The Black-throated Diver (*C. arcticus*) nests in the north and west of Scotland down to the shires of Perth and Argyll, in the Hebrides, and in the Orkneys, as well as abroad from Iceland, the Færoes, and Scandinavia eastward to the Pacific. It does not seem to occur to the north of the Arctic continent, but extends to Arctic America as a paler-naped form. It is distinguished from the Great Northern Diver by its grey crown and hind-neck, entirely black throat and chin, divided by a small band streaked with white, and by the similarly streaked sides of the neck. The habits of the two birds are alike, but the present species is rarer on our shores in winter.

The smallest and most common of our Divers is the Red-throated (*C. stellatus*), which breeds from Argyllshire up the west of Scotland and thence to

Red-throated Diver

Caithness, the Orkneys and Shetlands, as well as in north-west Ireland. In winter far larger numbers are found on our coasts than breed in our islands, which fact shews that there is an influx from abroad, where the bird ranges throughout the Arctic and Subarctic countries right round the Pole. In habits it resembles its congeners, while its harsh cries are supposed to portend storms, and the northern Scot knows it as the "Rain-goose." Less preference is shewn for islands as a site for the nest, which in the case of this species is often no nest at all, but merely a depression on the grassy side of a lake or in a "flow" on the northern moors; the eggs are comparatively small. The Redthroated Diver is brownish above, without distinct spots except in the young, and white below; the head and neck are pale grey, with white streaks on the former, while a long patch of red decorates the fore-neck. In winter this red is lost, and young birds, as in all our Divers, are entirely white below and brownish above, with pale margins to the feathers.

Family PODICIPEDIDÆ, or Grebes

The Great Crested Grebe (*Podicipes cristatus*) is a remarkable bird, brown above and glistening white below, which has the top of the head and its pair of backward tufts rich brown, while the pendant ruff is of a fine chestnut colour with black edge. These ornaments are lost at the autumn moult. The cheeks and a bar on each wing are white. Some years ago this species had become rare in Britain, except on the Norfolk Broads, but now it has increased to a most remarkable extent, and apart from the Highlands of Scotland there are few large lakes, reservoirs, or even smaller pieces of water with the requisite cover of reeds or tall sedge where it does not breed or may not be expected to do so in the near future. In winter it is often found on the sea, generally in or near some considerable estuary. The foreign range is extensive, as it includes central and southern Europe and Asia, reaching from the Baltic to Japan, Australia, New Zealand, India, and north Africa. The flight is fairly strong, but usually not long sustained; the food is of fishes, crustaceans, frogs, and so forth, with a little vegetable matter; the cry is croaking, with a short double alarm-note. The nest is a mass of aquatic plants placed in rather deep water or even floating; the four to six eggs are bluish white with a chalky crust and are covered with

Great Crested Grebe

water-weeds by the female on leaving them, if she is not too suddenly disturbed. Grebes often carry their downy young on their backs, or dive with them in that position.

The Red-necked Grebe (*P. griseigena*) is not uncommon on our shores from autumn to spring, though its numbers differ from year to year. It does not, however, breed with us, but in Scandinavia, Holland, the whole Baltic region, and across Russia to the Caspian Sea, as well as more rarely in other parts of Europe.

Somewhere about the middle of Asia it meets Holböll's Grebe, a larger and fairly distinguishable form ranging thence to Greenland and down North America. The Red-necked Grebe is smaller than the Great Crested Grebe and lacks the head ornaments, while the cheeks and throat are grey and the fore-neck reddish chestnut. The head, moreover, is blacker. The eggs are smaller.

The Horned or Slavonian Grebe (*P. auritus*) resembles its congeners in being mainly brown above and white below, but the black head and ruff are surmounted by two long chestnut tufts, one on each side of the head, while the breast and flanks are of a darker chestnut hue. It is often common on our coasts in the colder part of the year, especially to the northward, and has quite recently been found breeding in Invernessshire; it is not a southern species, but ranges from Iceland, Denmark, Scandinavia, and Russia to northern Asia and the northern part of North America.

The Black-necked Grebe (*P. nigricollis*) is about the same size as the last-named, both being smaller than the Red-necked species; it is less common on our coasts than the Horned Grebe, and occupies a comparatively southerly range, from about the Baltic to south Africa and temperate Asia generally. In the case of all of our Grebes several nests are often found in close proximity, but this species forms very regular colonies, though otherwise its habits are similar. It has the whole head and neck black, except for a patch of yellowish chestnut on each side of the face, and also chestnut flanks. It now breeds occasionally in Britain.

The Little Grebe (*P. fluviatilis*) is probably familiar to most of our readers as the "Dabchick," for it is fairly plentiful on our lakes, streams, and larger ponds, as well

Little Grebe's nest

Little Grebe's nest with eggs covered

as on the coast in winter-time. It breeds in temperate parts of the whole Palæarctic region. All Grebes being similar in general habits, those of the present species need not be given in detail; but it should be noted that it is an exceptionally wary little creature, diving at the shortest notice and staying a considerable time under water, or coming up only when it has reached the shelter of the sedge- or reed-beds. Thus it often escapes notice at its breeding haunts, unless many pairs are nesting in company. The note is a shrill reiterated "whit." The general colour is rather dark brown, with whitish under parts and ruddy chestnut cheeks and fore-neck.

ORDER XVII. TUBINARES

The Petrel group is distinguished from all other birds by the curious tubular nostrils, which lie exposed on the top of the bill, connected as a general rule, but widely separated in Albatroses. Moreover, we here find the greatest capacity for long-enduring flight known in the whole Class, though the Albatroses shew it to the greatest advantage. The bill varies from small and weak to immensely strong, the larger species having it formed of separate plates and often sharply hooked. The feet may be long or short, the front toes are fully webbed, the hind toe is small or absent. The wings are almost invariably long and pointed, the tail is moderate and in some cases forked. The downy nestlings vary in colour, but are often greyish white or brownish. Many of the species are chiefly nocturnal and stay in their holes during the day, but the Albatroses breed and live in the open. Petrels when handled throw up an evil-smelling oil, and the single egg has a more or less musky odour.

Order XVII. Tubinares

Family PROCELLARIIDÆ, or Petrels

The Storm-petrel (*Thalassidroma pelagica*), the Mother Carey's chicken of sailors, who believe that its presence portends a storm, is a black bird with white rump and vent-region. It flits lightly along the surface of the waves in mid-ocean, and does not confine itself to

Storm-Petrel

the coasts, while its feet commonly hang down and touch the water, so that it appears to be walking on it. The name "Petrel" has reference to this habit, and is derived from St Peter's attempt to do the same. This species breeds from the Scilly Islands and many parts of Ireland up the west of Britain to Shetland, but very rarely on our eastern or southern coasts; it is, however, very local in England, as it requires low rocky islands or at least wild rocky shores on which to breed. In fact, what are called in north and in west Britain "holms," that is, low flat islets covered with short grass, sea-pink and scattered boulders, are the true home of the bird with us. There it makes its nest, if any, of a little dry herbage in a burrow and lays its large roundish white egg with faint ruddy spots. Crevices among rocks and stones often take the place of a burrow, while the egg is seldom laid before June. Abroad the Storm-petrel breeds in the Færoes, Brittany (and the Channel Islands), the Mediterranean, and perhaps Scandinavia, but these limits must not be considered absolutely certain, as the bird may be confounded with other allied species. The food consists of small fishes and other sea creatures, with scraps obtained from passing ships; the note, seldom heard except in the burrow, is a sort of chirp constantly repeated. Petrels can with difficulty rise from the ground and at first fly feebly.

Leach's or the Fork-tailed Petrel (*Oceanodroma leucorrhoa*) has been proved, since 1847, to breed on St Kilda, North Rona, and other Hebridean Isles, as well as off the west of Ireland; its habits are identical with those of the last species, but the egg is a little larger. In colour the bird is slightly more grey, but the real distinction lies in the tail, which is deeply forked. We will not attempt to define its foreign range, but it is common on both the Atlantic and Pacific sides of North America down to the Bay of Fundy and California.

Family PUFFINIDÆ, or Shearwaters

The Great Shearwater (*Puffinus gravis*) is one of a different group of Petrels, much larger and heavier in their movements than those already described, which are therefore included in a separate family. Their flight is strong, but the birds are usually seen skimming along close to the waves or dashing down into the water, under which they easily follow their prey of fish, cuttle-fish, and other sea animals. The Great Shearwater is not very rare on our coasts, especially those that are rocky, and is represented by a paler form in the Mediterranean and the Atlantic Islands; but it is a curious fact that only one of the breeding-places of our form is known, viz. in Tristan da Cunha, and many persons believe that they all lie in the southern hemisphere. But the bird occurs certainly in the Atlantic Ocean from Greenland to the extreme south of America and Africa, so that we may hope for information on this point before long. Several Petrels breed both in the Atlantic and the Pacific, while their range is often hard to determine, as their presence does not always imply that they are breeding in the vicinity. Shearwaters are to a great extent nocturnal, and their weird crooning thrice-repeated cries are most commonly heard in the dusk. The upper parts of the Great Shearwater are brown, the lower white, while the neck is white all round and the upper tail-coverts are particoloured.

The Sooty and Dusky Shearwaters may be left to our list of irregular visitants; the former has been perpetually confounded with *P. gravis*, though it is a much darker bird, and there are two forms, or species, of the latter. In fact, modern ornithologists have subdivided this genus to such an extent that to determine any form recourse must be had to a long series of specimens. But we have one that breeds in Britain, with which we are more particularly concerned. This is the Manx Shearwater (*P. puffinus*), so called from its occurrence in the Isle of Man. There it breeds on the islet called the " Calf," while it is also found in the isles of Scilly, in those off the Welsh coast, in Ireland, and up the west of Scotland to Shetland, but not on our eastern shores. Thence the range extends to Iceland and the North Sea generally, the southern and Mediterranean species being the nearly allied Levantine Shearwater. The colour and habits resemble those of the Greater Shearwater, but our bird sits on the water and dives, perhaps to a greater extent than the larger form. The nest of grass and a little other material is at the end of a burrow, the site chosen, if not on small islands, being usually on the abrupt crumbling earthy slopes just above some high cliff, which often overhangs the sea. The large white egg has not so strong a smell of musk as that of the Fulmar or even the Storm-petrel, but the bird ejects an exceptional amount of evil-smelling oil when handled.

Our main list of British birds comes to an end with the Fulmar Petrel (*Fulmarus glacialis*), which until the beginning of last century was considered mainly an Arctic species, ranging from the north Atlantic eastwards along the northern coasts of Europe and perhaps Asia. The north Pacific species are different, though akin to our bird, and at present it is impossible to say where the forms meet or overlap in Asia. The Fulmar provides one of the most interesting illustrations of extension of range in the whole Class of Birds, for beginning from the ancient colony at St Kilda it has spread to the Shetlands and thence to the north of Scotland and even Ireland. Every year finds the birds pushing further southward. Each hen lays a big white egg, occasionally with a few small reddish spots, on a ledge of some precipice, for the most part well out of

Fulmars on eggs

reach without the aid of a rope; no nest is made, but an excavation is scraped in the bare earth or short turf, and there the bird may be seen sitting in company with many others of its species, while Guillemots, Razorbills, Puffins, Gulls, and Kittiwakes are scattered over the lower parts of the same rock-face. The note is low and of a crooning nature, the flight strong, but often performed in circles round the cliffs, when the birds have a very different appearance from Gulls, which they resemble in colour, as they float softly along, barely rising above the face of the rock and falling again to the level of their breeding-places. On the open sea this species is well known to sealers and whalers, whom it follows for the sake of the refuse and offal. It always settles on the water to feed, and its diet must doubtless consist mainly of fishes.

Occasional Visitors

List of occasional visitors to Britain, not discussed in the foregoing pages. Most of them are described in Howard Saunders' "Manual of British Birds."

Turdus musicus. Continental Song-Thrush.
Turdus fuscatus. Dusky Thrush.
Turdus atrigularis. Black-throated Thrush.
Turdus aureus. White's Thrush.
Turdus torquatus alpestris. Alpine Ring-Ousel.
Monticola saxatilis. Rock-Thrush.
Œnanthe œnanthe leucorrhoa. Greenland Wheatear.
Œnanthe isabellina. Isabelline Wheatear.
Œnanthe stapazina. Western Black-eared Wheatear.
Œnanthe stapazina amphileuca. Eastern Black-eared Wheatear.
Œnanthe occidentalis. Western Black-throated Wheatear.
Œnanthe deserti. Western Desert-Wheatear.
Œnanthe deserti albifrons. Eastern Desert-Wheatear.
Œnanthe leucomela. Pied Wheatear.
Œnanthe leucura. Black Wheatear.
Saxicola indica. Indian Stonechat.
Cyanosylvia suecica cyanecula. White-spotted Bluethroat.
Erithacus rubecula. Continental Robin.
Luscinia luscinia. Sprosser, or Eastern Nightingale.
Sylvia orphea. Orphean Warbler.
Sylvia melanocephala. Sardinian Warbler.
Sylvia nisoria. Barred Warbler.
Sylvia subalpina. Subalpine Warbler.
Phylloscopus fuscatus. Dusky Warbler.
Phylloscopus proregulus. Pallas's Warbler.
Phylloscopus superciliosus. Yellow-browed Warbler.
Phylloscopus viridanus. Greenish Warbler.
Phylloscopus collybita abietinus. Scandinavian Chiffchaff.
Phylloscopus tristis. Siberian Chiffchaff.
Phylloscopus trochilus eversmanni. Russian Willow-warbler.
Phylloscopus borealis. Arctic Willow-Warbler.
Hypolais icterina. Icterine Warbler.
Hypolais polyglotta. Melodious Warbler.

Occasional Visitors

Agrobates galactodes. Rufous Warbler.
Agrobates galactodes syriacus. Syrian Warbler.
Acrocephalus dumetorum. Blyth's Reed-Warbler.
Acrocephalus arundinaceus. Great Reed-Warbler.
Acrocephalus aquaticus. Aquatic Warbler.
Lusciniola schwartzi. Radde's Grasshopper-Warbler.
Locustella lanceolata. Temminck's Grasshopper-Warbler.
Locustella certhiola. Pallas's Grasshopper-Warbler.
Cettia cettii. Cetti's Warbler.
Accentor collaris. Alpine Accentor.
Cinclus cinclus. Black-bellied Dipper.
Ægithalus caudatus. White-headed Long-tailed Titmouse.
Parus ater. Continental Coal-Titmouse.
Parus borealis. Northern Willow-Titmouse.
Certhia familiaris. Northern Tree-Creeper.
Tichodroma muraria. Wall-Creeper.
Motacilla flava beema. Sykes' Yellow Wagtail.
Motacilla flava cinereicapilla. Ashy-headed Yellow Wagtail.
Motacilla flava thunbergi. Grey-headed Yellow Wagtail.
Motacilla feldeggi. Black-headed Wagtail.
Anthus cervinus. Red-throated Pipit.
Anthus campestris. Tawny Pipit.
Anthus richardi. Richard's Pipit.
Anthus spinoletta. Water-Pipit.
Anthus spinoletta rubescens. American Water-Pipit.
Anthus petrosus littoralis. Scandinavian Rock-Pipit.
Lanius meridionalis. Southern Grey Shrike.
Lanius senator badius. Corsican Woodchat Shrike.
Lanius nubicus. Masked Shrike.
Muscicapa latirostris. Brown Flycatcher.
Muscicapa collaris. White-collared Flycatcher.
Muscicapa parva. Red-breasted Flycatcher.
Hirundo rufula. Red-rumped Swallow.
Carduelis carduelis. Continental Goldfinch.
Spinus citrinella. Citril Finch.
Serinus serinus. Serin Finch.
Montifringilla nivalis. Snow-Finch.
Acanthis linaria rostrata. Greenland Redpoll.
Acanthis linaria holbölli. Holböll's Redpoll.

Occasional Visitors

Acanthis hornemanni. Hornemann's Redpoll.
Acanthis hornemanni exilipes. Hoary Redpoll.
Carpodacus erythrinus. Rosy Bullfinch.
Pyrrhula pyrrhula. Northern Bullfinch.
Pinicola enucleator. Pine-Grosbeak.
Loxia leucoptera. White-winged Crossbill.
Loxia leucoptera bifasciata. Two-barred Crossbill.
Emberiza melanocephala. Black-headed Bunting.
Emberiza leucocephala. Pine-Bunting.
Emberiza cia. Meadow-Bunting.
Emberiza hortulana. Ortolan Bunting.
Emberiza cioides castaneiceps. East Siberian Meadow-Bunting.
Emberiza aureola. Yellow-breasted Bunting.
Emberiza rustica. Rustic Bunting.
Emberiza pusilla. Little Bunting.
Emberiza palustris. Large-billed Reed-Bunting.
Emberiza palustris tschusii. Eastern Large-billed Reed-Bunting
Nucifraga caryocatactes. Thick-billed Nutcracker.
Nucifraga caryocatactes macrorhynchus. Thin-billed Nutcracker.
Alauda arvensis cinerascens. Eastern Skylark.
Calandrella brachydactyla. Short-toed Lark.
Melanocorypha sibirica. White-winged Lark.
Melanocorypha yeltoniensis. Black Lark.
Micropus melba. Alpine or White-bellied Swift.
Chœtura caudacuta. Needle-tailed Swift.
Caprimulgus ruficollis. Red-necked Nightjar.
Caprimulgus ægyptius. Egyptian Nightjar.
Coracias garrulus. Roller.
Merops apiaster. Bee-eater.
Clamator glandarius. Great Spotted Cuckoo.
Coccyzus americanus. Yellow-billed Cuckoo.
Coccyzus erythrophthalmus. Black-billed Cuckoo.
Nyctea nyctea. Snowy Owl.
Surnia ulula. European Hawk-Owl
Surnia ulula caparoch. American Hawk-Owl.
Nyctala funerea. Tengmalm's Owl.
Otus scops. Scops Owl.
Bubo ignavus. Eagle Owl.
Gyps fulvus. Griffon Vulture.

Occasional Visitors

Neophron percnopterus. Egyptian Vulture.
Aquila maculata. Spotted Eagle.
Milvus migrans. Black Kite.
Falco peregrinus anatum. American Peregrine.
Falco vespertinus. Red-footed Falcon.
Falco naumanni. Lesser Kestrel.
Egretta alba. Great White Heron.
Egretta garzetta. Little Egret.
Ardeola ibis. Buff-backed Heron.
Nycticorax nycticorax. Night-Heron.
Butorides virescens. Little Green Heron.
Botaurus lentiginosus. American Bittern.
Ciconia nigra. Black Stork.
Plegadis falcinellus. Glossy Ibis.
Phœnicopterus antiquorum. Flamingo.
Anser erythropus. Lesser Pink-footed Goose.
Chen hyperboreus. Cassin's Snow-Goose.
Chen hyperboreus nivalis. Snow-Goose.
Branta ruficollis. Red-breasted Goose.
Tadorna casarca. Ruddy Sheldrake.
Mareca americana. American Wigeon.
Querquedula discors. Blue-winged Teal.
Querquedula crecca carolinensis. American Green-winged Teal.
Netta rufina. Red-crested Pochard.
Nyroca nyroca. White-eyed Duck.
Glaucion islandica. Barrow's Golden-eye.
Glaucion albeola. Buffel-headed Duck.
Histrionicus histrionicus. Harlequin-Duck.
Heniconetta stelleri. Steller's Duck.
Somateria spectabilis. King-Eider.
Œdemia perspicillata. Surf-Scoter.
Lophodytes cucullatus. Hooded Merganser.
Streptopelia orientalis. Eastern Turtle-dove.
Turnix sylvatica. Andalusian Hemipode.
Porzana carolina. Carolina Rail.
Houbara undulata macqueeni. Macqueen's Bustard.
Glareola pratincola. Collared Pratincole.
Glareola nordmanni. Black-winged Pratincole.
Cursorius gallicus. Cream-coloured Courser.

Occasional Visitors

Charadrius dominicus. American Golden Plover.
Charadrius dominicus fulvus. Eastern Golden Plover.
Ægialitis dubia. Little Ringed Plover.
Ægialitis vocifera. Killdeer Plover.
Ægialitis asiatica. Caspian Plover.
Chettusia gregaria. Sociable Plover.
Himantopus himantopus. Black-winged Stilt.
Limicola platyrhyncha. Broad-billed Sandpiper.
Terekia cinerea. Terek Sandpiper.
Tringa maculata. American Pectoral Sandpiper.
Tringa acuminata. Siberian Pectoral Sandpiper.
Tringa fuscicollis. Bonaparte's Sandpiper.
Tringa bairdi. Baird's Sandpiper.
Tringa minutilla. American Stint.
Ereunetes pusillus. Semipalmated Sandpiper.
Tryngites subruficollis. Buff-breasted Sandpiper.
Bartramia longicauda. Bartram's Sandpiper.
Totanus macularius. Spotted Sandpiper.
Totanus solitarius. Solitary Sandpiper.
Totanus stagnatilis. Marsh-Sandpiper.
Totanus melanoleucus. Greater Yellowshank.
Totanus flavipes. Yellowshank.
Macrorhamphus griseus. Red-breasted Snipe.
Numenius borealis. Esquimaux Curlew.
Numenius tenuirostris. Slender-billed Curlew.
Sterna caspia. Caspian Tern.
Sterna anglica. Gull-billed Tern.
Sterna fuliginosa. Sooty Tern.
Anous stolidus. Noddy Tern.
Pagophila eburnea. Ivory Gull.
Larus cachinnans. Yellow-legged Herring-Gull.
Larus ichthyaëtus. Great Black-headed Gull.
Larus melanocephalus. Adriatic or Mediterranean Gull.
Larus philadelphia. Bonaparte's Gull.
Rhodostethia rosea. Cuneate-tailed Gull.
Oceanites oceanicus. Wilson's Petrel.
Pelagodroma marina. Frigate Petrel.
Oceanodroma castro. Madeiran Petrel.
Puffinus puffinus yelkouanus. Levantine Shearwater.

Puffinus griseus. Sooty Shearwater.
Puffinus obscurus baroli. Little Madeiran Shearwater.
Puffinus kuhli. Mediterranean Great Shearwater.
Œstrelata hasitata. Capped Petrel.
Œstrelata brevipes. Collared Petrel.
Œstrelata neglecta. Kermadec Petrel.
Bulweria bulweri. Bulwer's Petrel.
Diomedea melanophrys. Black-browed Albatros.

INDEX

Acanthis cabaret, 76
Acanthis cannabina, 75
Acanthis flavirostris, 76
Acanthis hornemanni, 259
Acanthis hornemanni exilipes, 259
Acanthis linaria, 77
Acanthis linaria holbölli, 258
Acanthis linaria rostrata, 258
Accentor, Alpine, 258
Accentor collaris, 258
Accentor modularis, 39
Accentorinæ, 15, 39
Accipiter nisus, 123
Acrocephalus aquaticus, 258
Acrocephalus arundinaceus, 258
Acrocephalus dumetorum, 258
Acrocephalus palustris, 34
Acrocephalus schœnobænus, 34
Acrocephalus streperus, 34
Ægialitis alexandrina, 192
Ægialitis asiatica, 261
Ægialitis dubia, 261
Ægialitis hiaticula, 191
Ægialitis vocifera, 261
Ægithalus caudatus, 44, 258
Ægithalus caudatus roseus, 44
Agrobates galactodes, 258
Agrobates galactodes syriacus, 253
Alauda arvensis, 95
Alauda arvensis cinerascens, 259
Alaudidæ, 15, 95
Albatros, 6, 250
Albatros, Black-browed, 262
Alca impennis, 236
Alca torda, 238
Alcæ. 17, 236
Alcedinidæ, 16, 107
Alcedo ispida, 107
Alcidæ, 17, 236
Alcinæ, 17, 236
Alle alle, 241

Ampelidæ, 15, 63
Ampelis garrulus, 63
Anas boschas, 150
Anas strepera, 151
Anatidæ, 16, 143, 144
Anatinæ, 16, 149
Anous stolidus, 261
Anser albifrons, 146
Anser anser, 144
Anser brachyrhynchus, 146
Anser erythropus, 260
Anser fabalis, 146
Anseres, 16, 143
Anserinæ, 16, 144
Anthus bertheloti, 59
Anthus campestris, 258
Anthus cervinus, 258
Anthus petrosus, 60
Anthus petrosus littoralis, 256
Anthus pratensis, 58
Anthus richardi, 258
Anthus spinoletta, 258
Anthus spinoletta rubescens, 258
Apteria, 3
Aquila chrysaëtus, 121
Aquila maculata, 250
Archæopteryx, 1
Ardea cinerea, 136
Ardea purpurea, 138
Ardeidæ, 16, 136
Ardeola ibis, 260
Ardeola ralloides, 139
Arenaria interpres, 195
Asio accipitrinus, 113
Asio otus, 112
Astur palumbarius, 123
Auk, 236
Auk, Great, 236, 237
Auk, Little, 241
Aves (Class), 1
Avocet, 188, 197, 198

264 Index

Barbicels, 4
Barbs, 3
Barbules, 3
Bargander, 149
Bartramia longicauda, 261
Bass Goose, 136
Beam-bird, 65
Bee-eater, 99, 259
Birds in general, 1
Bittern, 136, 139, 140
Bittern, American, 260
Bittern, Little, 139, 140
Blackbird, 20
Blackbird, Hill-, 20
Blackcap, 28, 30, 31
Blue Darr, 221
Blue Rock, 165
Blue-throat, 22
Blue-throat, Red-spotted, 26
Blue-throat, White-spotted, 257
Bonxie, 233
Botaurus lentiginosus, 260
Botaurus stellaris, 140
Bottle-tit, 46
Brambling, 74
Branta bernicla, 147
Branta leucopsis, 147
Branta ruficollis, 260
Bubo ignavus, 259
Bullfinch, 77, 79
Bullfinch, Northern, 259
Bullfinch, Rosy, 259
Bulweria bulweri, 262
Bunting, 69, 80, 81
Bunting, Black-headed, 259
Bunting, Cirl, 81
Bunting, Corn-, 80, 82
Bunting, East Siberian Meadow-, 259
Bunting, Eastern Large-billed Reed-, 259
Bunting, Lapland, 83
Bunting, Large-billed Reed-, 259
Bunting, Little, 259
Bunting, Meadow-, 259
Bunting, Ortolan, 259
Bunting, Pine-, 259
Bunting, Reed-, 82
Bunting, Rustic, 259
Bunting, Snow-, 13, 84
Bunting, Yellow, 81
Bunting, Yellow-breasted, 259

Bustard, Great, 184, 186
Bustard, Little, 187
Bustard, Macqueen's, 260
Butcher-bird, 61
Buteo buteo, 120
Buteo lagopus, 120
Butorides virescens, 260
Buzzard, African, 120
Buzzard, Common, 120, 124
Buzzard, Honey-, 125
"Buzzard," Moor-, 117
Buzzard, Rough-legged, 120

Caccabis petrosa, 176
Caccabis rufa, 176
Caccabis saxatilis, 176
Calandrella brachydactyla, 259
Calcarius lapponicus, 83
Capercaillie, 170, 172
Caprimulgidæ, 16, 101
Caprimulgus ægyptius, 259
Caprimulgus europæus, 101
Caprimulgus ruficollis, 259
Carduelis carduelis, 258
Carduelis carduelis britannica, 71
Carinatæ, 3, 15
Carine noctua, 115
Carpodacus erythrinus, 259
Carr-Swallow, 221
Cassowary, 3
Catharacta skua, 232
Certhia familiaris, 54, 258
Certhia familiaris britannica, 54
Certhiidæ, 15, 54
Cettia cettii, 258
Chætura caudacuta, 259
Chaffinch, 74, 75
Charadriidæ, 17, 190
Charadrius apricarius, 192
Charadrius dominicus, 261
Charadrius dominicus fulvus, 261
Chat, 22
Chen hyperboreus, 260
Chen hyperboreus nivalis, 260
Chettusia gregaria, 261
Chiffchaff, 33, 34, 46
Chiffchaff, Scandinavian, 36, 257
Chiffchaff, Siberian, 33, 257
Chloris chloris, 70
Chough, 87
Ciconia ciconia, 141
Ciconia nigra, 260

Index

Ciconiidæ, 16, 141
Cinclidæ, 15, 39
Cinclus cinclus, 39, 258
Cinclus cinclus britannicus, 40
Circus æruginosus, 117
Circus cyaneus, 119
Circus pygargus, 119
Clamator glandarius, 259
Clangula hyemalis, 158
Classification, 15
Cobbler's Awl, 198
Coccothraustes coccothraustes, 70
Coccyzus americanus, 259
Coccyzus erythrophthalmus, 259
Columba livia, 165
Columba œnas, 165
Columba palumbus, 164
Columbæ, 16, 163
Columbidæ, 16, 164
Colymbidæ, 17, 243
Colymbus adamsi, 244
Colymbus arcticus, 244
Colymbus immer, 243
Colymbus stellatus, 245
Coot, 179, 182–184, 188
Coracias garrulus, 259
Coraciidæ, 16
Cormorant, 132–135
Cormorant, Green, 134
Corvidæ, 15, 87
Corvus, 90
Corvus corax, 91
Corvus cornix, 93
Corvus corone, 93
Corvus frugilegus, 94
Corvus monedula, 90
Coturnix coturnix, 178
Courser, Cream-coloured, 260
Crake, Baillon's, 181
Crake, Corn-, 179
Crake, Little, 181
Crake, Spotted, 181, 182
Crane, 184, 185
Creeper, Northern Tree-, 258
Creeper, Tree-, 54
Creeper, Wall-, 258
Crex crex, 179
Crossbill, 79, 80
Crossbill, Parrot-, 80
Crossbill, Two-barred, 259
Crossbill, White-winged, 259
Crow, 90, 93, 95, 113

Crow, Carrion-, 93
Crow, Grey, 93
Crow, Hooded, 93
Crow tribe, 87, 90
"Crow," Water-, 40
Cuckoo, 99, 109–111
Cuckoo, Black-billed, 259
Cuckoo, Great Spotted, 259
Cuckoo, Yellow-billed, 259
Cuckoo's Mate, 102
Cuculidæ, 16, 109
Cuculus canorus, 109
Curlew, 216–218
Curlew, Esquimaux, 261
Curlew, Slender-billed, 261
Cursorius gallicus, 260
Cyanosylvia suecica, 26
Cyanosylvia suecica cyanecula, 257
Cygninæ, 16, 147
Cygnus bewicki, 148
Cygnus cygnus, 148
Cygnus olor, 148
Cypselidæ, 15, 99

Dabchick, 248
Dafila acuta, 152
Decorative plumes, 5
Delichon urbica, 67
Deviling, 100
Diomedea melanophrys, 262
Diomedeidæ, 17
Dipper, 39
Dipper, Black-billed, 258
Dishwasher, 56
Diver, 243, 244
Diver, Black-throated, 244
Diver, Great Northern, 243, 244
Diver, Red-throated, 245, 246
Dotterel, 190
Dove, 163, 164
Dove, Ring-, 164, 165
Dove, Rock-, 165, 168
Dove, Stock-, 165, 166
Dove, Turtle-, 167
Down of nestling, 3
Dryobates major, 103
Dryobates minor, 105
Duck, 143, 144, 148, 149
Duck, Barrow's Golden-eye, 260
Duck, Buffel-headed, 260
Duck, Eider, 159, 160
Duck, Golden-eye, 158

Duck, Harlequin-, 260
Duck, Long-tailed, 144, 158
Duck, Pintail, 144, 152
Duck, Scaup, 157
Duck, Sea-, 156
Duck, Steller's, 260
Duck, Tufted, 156, 157
Duck, White-eyed, 260
Duck, Wild, 150
Ducks, Fresh-water, 149
Dunbird, 156
Dunlin, 204

Eagle, 117
Eagle, Spotted, 260
Eagle, White-tailed, 122, 123
Eclipse Plumage, 5
Egret, Little, 260
Egretta alba, 260
Egretta garzetta, 260
Eider, 159, 161
Eider, King-, 144, 260
Emberiza aureola, 259
Emberiza calandra, 80
Emberiza cia, 259
Emberiza cioides castaneiceps, 259
Emberiza cirlus, 82
Emberiza citrinella, 81
Emberiza hortulana, 259
Emberiza leucocephala, 259
Emberiza melanocephala, 259
Emberiza palustris, 259
Emberiza palustris tschusii, 259
Emberiza pusilla, 259
Emberiza rustica, 259
Emberiza schœniclus, 82
Emberizinæ, 15, 69, 80
Emu, 3
Ereunetes pusillus, 261
Erithacus rubecula, 27, 257
Eudromias morinellus, 190

Falco æsalon, 129
Falco naumanni, 260
Falco peregrinus, 126
Falco peregrinus anatum, 260
Falco subbuteo, 128
Falco tinnunculus, 130
Falco vespertinus, 260
Falcon, Greenland, 125
Falcon, Gyr, 126
Falcon, Iceland, 125

Falcon, Red-footed, 260
Falconidæ, 16, 117
Families of Birds, 15
Feather, 3
Feather-poke, 46
Feltyfare, 19
Fieldfare, 19, 21, 22
Finch, 69
Finch, Citril, 258
Finch, Gold-, see Goldfinch
Finch, Green-, see Greenfinch
Finch, Serin, 258
Finch, Snow-, 258
Fire-crested "Wren," 50, 51
Firetail, 25
Flamingo, 260
Flammea flammea, 112
Flight, 6–8
Flycatcher, 64
Flycatcher, Brown, 258
Flycatcher, Pied, 66
Flycatcher, Red-breasted, 258
Flycatcher, Spotted, 64
Flycatcher, White-collared, 258
Fowl, 169
Fratercula arctica, 241
Fraterculinæ, 17, 241
Fringilla cœlebs, 74
Fringilla montifringilla, 74
Fringillidæ, 15, 69
Fringillinæ, 15, 69, 70
Fulica atra, 184
Fulicariæ, 17, 179
Fuligulinæ, 16, 156
Fulmar, see Petrel
Fulmarus glacialis, 254, 255

Gadwall, 151
Galerida cristata, 98
Gallinæ, 17, 169
Gallinago gallinago, 202
Gallinago media, 203
Gallinula chloropus, 182
Gannet, 5, 132, 134, 135
Garefowl, 236
Garganey, 153
Garrulus glandarius, 89
Gaviæ, 17, 219
Geese, 143, 144
Geese, Black, 144, 147
Geese, Grey, 144, 146, 147
Glareola nordmanni, 260

Index

Glareola pratincola, 260
Glareolidæ, 17
Glaucion albeola, 260
Glaucion clangula, 158
Glaucion islandica, 260
Gloss, 6
Goatsucker, 99, 101, 102
Godwit, Bar-tailed, 215, 216
Godwit, Black-tailed, 216
Gold-crested "Wren," 50
Golden Eagle, 121, 213
Golden-eye, see Duck
Golden-eye, Barrow's, 260
Goldfinch, 71, 72
Goldfinch, Continental, 258
Goosander, 161, 162
Goose, Bean-, 146, 147
Goose, Bernacle, 147
Goose, Brent, 147
Goose, Cassin's Snow-, 260
Goose, Gambel's, 146
Goose, Grey Lag, 144–146
Goose, Lesser Pink-footed, 260
Goose, Lesser White-fronted, 146
Goose, Pink-footed, 146, 147
Goose, Red-breasted, 260
Goose, Snow-, 260
Goose, White-fronted, 146
Goshawk, 123, 124
Grallæ, 17, 179
Grebe, 243, 246
Grebe, Black-necked, 248
Grebe, Great Crested, 246, 248
Grebe, Holböll's, 248
Grebe, Horned, 248
Grebe, Little, 248, 249
Grebe, Red-necked, 247, 248
Grebe, Slavonian, 248
Greenfinch, 70, 76, 79
Greenshank, 214–215
Grosbeak, Pine-, 259
Grouse, 167, 169
Grouse, Black, 170, 171
Grouse, Red, 13, 169, 170, 173
"Grouse," Sand-, 164, 168
Grouse, Willow-, 172, 173
Grues, 17, 184
Gruidæ, 184
Grus grus, 184
Guillemot, 236, 238
Guillemot, Black, 236, 240
Guillemot, Brünnich's, 240

Guillemot, Common, 238–240
Guillemot, Ringed, 240
Gull, 219–221, 225
Gull, Adriatic, 261
Gull, Black-headed, 220, 226
Gull, Bonaparte's, 227, 261
Gull, Common, 220, 227, 228, 230
Gull, Cuneate-tailed, 261
Gull, Glaucus, 230
Gull, Great Black-backed, 220, 229, 230
Gull, Great Black-headed, 226, 261
Gull, Herring-, 228, 229
Gull, Iceland, 230
Gull, Ivory, 261
Gull, Lesser Black-backed, 228–230
Gull, Little, 226, 227
Gull, Mediterranean, 261
Gull, Parasitic, 233
Gull, Sabine's, 219, 225, 227
Gull-teaser, 233
Gull, Yellow-legged Herring-, 228, 261
Gyps fulvus, 259

Hæmatopus ostralegus, 196
Haliaëtus albicilla, 122
Harrier, 117–119
Harrier, Hen-, 119
Harrier, Marsh-, 117
Harrier, Montagu's, 119
Hawfinch, 70, 71
Hawk, Fish-, 132
Hawk, Hunting, 126
Hawk, Sparrow-, 123
Heather Lintie, 76
Hedgesparrow, 39, 111
Hemipode, Andalusian, 260
Hemipodes, 5
Heniconetta stelleri, 260
Hern, 136
Herodiones, 16, 136
Heron, 136, 138, 186
Heron, Buff-backed, 260
Heron, Great White, 260
Heron, Little Green, 260
Heron, Night-, 139, 260
Heron, Purple, 138
Heron, Squacco, 139
Hierofalco candicans, 125
Hierofalco gyrfalco, 126

Hierofalco islandus, 125
Himantopus himantopus, 261
Hirundinidæ, 66
Hirundo rufula, 258
Hirundo rustica, 66
Histrionicus histrionicus, 260
Hobby, 128
Hoopoe, 99, 108
Houbara undulata macqueeni, 260
Humming-bird, 67, 99, 101
Hydrochelidon leucopareia, 222
Hydrochelidon leucoptera, 222
Hydrochelidon nigra, 220
Hypolais icterina, 257
Hypolais polyglotta, 257

Ibididæ, 16
Ibis, 136
Ibis, Glossy, 142, 260
Iridescence, 6
Ixobrychus minutus, 139
Iynginæ, 16, 102
Iynx torquilla, 102

Jackdaw, 90
Jay, 89

Kestrel, 128, 130
Kestrel, Lesser, 260
Kingfisher, 99, 107, 108
"Kite," 119
Kite, 124
Kite, Black, 260
Kittiwake, 227, 230–234
Knot, 207, 208

Lagopus mutus, 173
Lagopus scoticus, 171
Laniidæ, 15, 61
Lanius collurio, 61
Lanius excubitor, 63
Lanius major, 63
Lanius meridionalis, 258
Lanius minor, 63
Lanius nubicus, 258
Lanius senator, 62
Lanius senator badius, 258
Lapwing, 194, 195
Laridæ, 17, 219
Larinæ, 17, 219, 225,
Lark, 55
Lark, Black, 259

Lark, Crested, 96, 98
Lark, Shore-, 98
Lark, Short-toed, 259
Lark, Sky-, 95, 97
Lark, White-winged, 259
Lark, Wood-, 60, 97
Larus argentatus, 228
Larus cachinnans, 261
Larus canus, 227
Larus fuscus, 228
Larus glaucus, 230
Larus ichthyaëtus, 261
Larus leucopterus, 230
Larus marinus, 229
Larus melanocephalus, 261
Larus minutus, 226
Larus philadelphia, 261
Larus ridibundus, 226
Lighthouse Observations, 9
Limicola platyrhyncha, 261
Limicolæ, 17, 188
Limnocryptes gallinula, 203
Limosa lapponica, 215
Limosa limosa, 216
Linnet, 70, 75
Linnet, Brown, 75
"Linnet," Green, 70
Linnet, Grey, 75
Linnet, Mountain-, 76
Linnet, Red, 75
Locustella certhiola, 258
Locustella lanceolata, 258
Locustella luscinioides, 38
Locustella nævia, 36
Loom, 238
Lophodytes cucullatus, 260
Loxia leucoptera bifasciata, 259
Loxia curvirostra, 79
Loxia curvirostra scotica, 79
Loxia leucoptera, 259
Lullula arborea, 97
Luscinia luscinia, 257
Luscinia megarhyncha, 28
Lusciniola schwartzi, 258
Lyrurus tetrix, 171

Machetes pugnax, 209
Macrorhamphus griseus, 261
Magpie, 89, 90, 113
Mallard, 150, 156
Mareca americana, 260
Mareca penelope, 155

Martin, House-, 67–69
Martin, Sand-, 67, 68
Melanocorypha sibirica, 259
Melanocorypha yeltoniensis, 259
Melizophilus undatus, 31
Merganser, 143
Merganser, Hooded, 260
Merganser, Red-breasted, 162
Mergellus albellus, 163
Merginæ, 16, 161
Mergus merganser, 161
Mergus serrator, 162
Merlin, 129, 130
Merops apiaster, 259
Micropus apus, 100
Micropus melba, 259
Migrants, Partial, 8
Migrants, True or Summer, 8
Migration, 8–13
Milvus migrans, 260
Milvus milvus, 124
Mistle-thrush, 19
Monticola saxatilis, 257
Montifringilla nivalis, 258
Moorhen, 182
Motacilla alba, 57
Motacilla boarula, 57
Motacilla feldeggi, 258
Motacilla flava, 58
Motacilla flava beema, 258
Motacilla flava cinereicapilla, 258
Motacilla flava thunbergi, 258
Motacilla lugubris, 56
Motacilla raii, 58
Motacillidæ, 15, 55
Mother Carey's Chicken, 251
Moult, 5
Murre, 238
Muscicapa atricapilla, 66
Muscicapa collaris, 258
Muscicapa grisola, 64
Muscicapa latirostris, 258
Muscicapa parva, 258
Muscicapidæ, 15, 64
Mussel-picker, 196

Neophron percnopterus, 260
Netta rufina, 260
Nidicolous, 6
Nidifugous, 6
Nightingale, 13, 22, 28
Nightingale, Eastern, 257

Nightjar, 101, 102
Nightjar, Egyptian, 259
Nightjar, Red-necked, 259
Nomenclature, 15
Nucifraga caryocatactes, 259
Nucifraga caryocatactes macrorhynchus, 259
Numenius arquata, 216
Numenius borealis, 261
Numenius phæopus, 218
Numenius tenuirostris, 261
Nutcracker, Thick-billed, 259
Nutcracker, Thin-billed, 259
Nuthatch, 51
Nutjobber, 51
Nyctala funerea, 259
Nyctea nyctea, 259
Nycticorax nycticorax, 139, 260
Nyroca ferina, 156
Nyroca fuligula, 157
Nyroca marila, 157
Nyroca nyroca, 260

Oceanites oceanicus, 261
Oceanodroma castro, 261
Oceanodroma leucorrhoa, 252
Œdemia fusca, 160
Œdemia nigra, 160
Œdemia perspicillata, 161, 260
Œdicnemidæ, 17, 188
Œdicnemus œdicnemus, 188
Œnanthe deserti, 257
Œnanthe deserti albifrons, 257
Œnanthe isabellina, 257
Œnanthe leucomela, 257
Œnanthe leucura, 257
Œnanthe occidentalis, 257
Œnanthe œnanthe, 22
Œnanthe œnanthe leucorrhoa, 257
Œnanthe stapazina, 257
Œnanthe stapazina amphileuca, 257
Œstrelata brevipes, 262
Œstrelata hasitata, 262
Œstrelata neglecta, 262
Old Squaw, 159
Orders of Birds, 15
Oriole, Golden, 60
Oriolidæ, 15, 60
Oriolus oriolus, 60
Osprey, 117, 131
Ostrich, 3

Otides, 17, 186
Otididæ, 17, 186
Otis tarda, 184
Otis tetrax, 187
Otocorys alpestris, 98
Otus scops, 259
Ousel, Alpine Ring-, 257
Ousel, Ring-, 20
Ousel, Water-, 40
Owl, 111, 112
Owl, American Hawk-, 259
Owl, Barn-, 112
Owl, Eagle-, 259
Owl, European Hawk-, 259
Owl, Little, 112, 115, 116
Owl, Long-eared, 112, 115
Owl, Scops, 259
Owl, Screech, 112
Owl, Short-eared, 113
Owl, Snowy, 259
Owl, Tawny, 114
Owl, Tengmalm's, 259
Owl, Wood-, 114
Owl, Woodcock-, 113
Oxbird, 205
Oxeye, 46
Oyster-catcher, 188, 196, 197, 207

Pagophila eburnea, 261
Pallas's Sand-grouse, 168
Pandion haliaëtus, 131
Panuridæ, 15, 42
Panurus biarmicus, 42
Paridæ, 15, 43
Partridge, 169, 174, 176, 177
Partridge, Barbary, 176
Partridge, Red-legged, 176
Parus, 43
Parus ater, 47, 258
Parus ater britannicus, 47
Parus borealis, 258
Parus borealis kleinschmidti, 48
Parus cæruleus, 49
Parus cristatus, 49
Parus major, 46
Parus palustris dresseri, 48
Passage, Birds of, 8
Passer domesticus, 73
Passer montanus, 73
Passeres, 15, 18
Pastor roseus, 86
Peacock, 169

Peewit, 188, 190, 194, 195
Pelagodroma marina, 261
Pelecanidæ, 16, 133
Pelican, 132
Penguins, 5, 11, 236
Perdix perdix, 176
Peregrine Falcon, 124, 126–128
Peregrine Falcon, American, 260
Pernis apivorus, 125
Petrel, 250–253
Petrel, Bulwer's, 262
Petrel, Capped, 262
Petrel, Collared, 262
Petrel, Fork-tailed, 252
Petrel, Frigate-, 261
Petrel, Fulmar, 254
Petrel, Kermadec, 262
Petrel, Leach's, 252
Petrel, Madeiran, 261
Petrel, Storm-, 251, 252
Petrel, Wilson's, 261
Phalacrocorax carbo, 133
Phalacrocorax graculus, 134
Phalarope, 5, 188
Phalarope, Grey, 199
Phalarope, Red-necked, 199
Phalaropus fulicarius, 199
Phalaropus lobatus, 199
Phasianidæ, 17, 174
Phasianus colchicus, 174
Pheasant, 169, 170, 176
"Pheasant," Reed-, 42
Phœnicopteridæ, 16
Phœnicopterus antiquorum, 260
Phœnicurus phœnicurus, 25
Phœnicurus titys, 26
Phylloscopus borealis, 257
Phylloscopus collybita, 33
Phylloscopus collybita abietinus, 257
Phylloscopus fuscatus, 257
Phylloscopus proregulus, 257
Phylloscopus sibilatrix, 33
Phylloscopus superciliosus, 257
Phylloscopus tristis, 257
Phylloscopus trochilus, 32
Phylloscopus trochilus eversmanni, 257
Phylloscopus viridanus, 257
Pica pica, 89
Picariæ, 16, 98, 99
Picidæ, 16, 102

Picinæ, 16, 103
Picus, 99
Picus viridis, 105
Pigeon, 163, 164, 167, 171
Pinicola enucleator, 259
Pintail, see Duck
Pipit, 55, 56, 58
Pipit, American Water-, 258
Pipit, Meadow-, 58, 60, 110
Pipit, Red-throated, 258
Pipit, Richard's, 258
Pipit, Rock-, 60
Pipit, Scandinavian Rock-, 258
Pipit, Tawny, 258
Pipit, Tree-, 59, 60, 97
Pipit, Water-, 258
Platalea leucorodia, 142
Plataleidæ, 16, 142
Plectrophenax nivalis, 84
Plegadis falcinellus, 260
Plover, 188, 190
Plover, American Golden, 261
Plover, Caspian, 261
Plover, Eastern Golden, 261
Plover, Golden, 168, 189, 192, 193
Plover, Grey, 193
Plover, Kentish, 192
Plover, Killdeer, 261
Plover, Little Ringed, 192, 261
Plover, Norfolk, 188
Plover, Ringed, 191, 192
Plover, Sociable, 261
Plover tribe, 167
Plover's Page, 205
Pochard, 156–158
Pochard, Red-crested, 260
Podicipedidæ, 17, 243, 246
Podicipes auritus, 248
Podicipes cristatus, 246
Podicipes fluviatilis, 248
Podicipes griseigena, 247
Podicipes nigricollis, 248
Porzana carolina, 260
Porzana parva, 181
Porzana porzana, 181
Porzana pusilla, 181
Post-bird, 65
Pratincole, Black-winged, 260
Pratincole, Collared, 260
Procellariidæ, 17, 251
Ptarmigan, 169, 170, 173, 174
Ptarmigan, Spitsbergen, 11

Pterocletes, 16, 167
Pteroclidæ, 16, 168
Pterylæ, 3
Puffin, 236, 241, 242
Puffinidæ, 17, 253
Puffinus gravis, 253
Puffinus griseus, 262
Puffinus kuhli, 262
Puffinus obscurus baroli, 262
Puffinus puffinus, 254
Puffinus puffinus yelkouanus, 261
Purre, 205
Pygopodes, 17, 243
Pyrrhocorax pyrrhocorax, 87
Pyrrhula pyrrhula, 77, 79, 259
Pyrrhula pyrrhula pileata, 79

Quail, 178
Quail. Button-, 178
Querquedula crecca, 153
Querquedula crecca carolinensis, 260
Querquedula discors, 260
Querquedula querquedula, 153

Rail, 179
Rail, Carolina, 260
Rail, Land-, 179
Rail, Water-, 179, 180, 181
Rainbird, 105
Rain-goose, 246
Rallidæ, 17, 179
Rallus aquaticus, 179
Raptorial Birds, 117
Ratitæ, 3
Raven, 91, 92
Razorbill, 238, 239
Recurvirostra avocetta, 197
Redbreast, 22, 27
Redpoll, Greenland, 258
Redpoll, Hoary, 259
Redpoll, Holböll's, 258
Redpoll, Hornemann's, 259
Redpoll, Lesser, 76, 77
Redpoll, Mealy, 76, 77
Redshank, 212–215
Redshank, Dusky, 214
Redshank, Spotted, 214
Redstart, 22, 25, 66
Redstart, Black, 26
Redwing, 21
Reeler, Fen-, 36

Reeve, 209, 210
Region, Australian, 13
Region, Ethiopian, 13
Region, Holarctic, 13
Region, Indian, 13
Region, Nearctic, 13
Region, Neotropical, 13
Region, Palæarctic, 13
Regions, Zoo-geographical, 13, 14
Regulidæ, 15, 50
Regulus ignicapillus, 50
Regulus regulus, 50
Residents, 9
Rhodostethia rosea, 261
Ringing of Birds, 9
Riparia riparia, 67
Rissa tridactyla, 230
Robin, 27, 111
Robin, Continental, 257
Roller, 99, 253
Rook, 93, 94
Ruff, 209, 210

Sanderling, 12, 188, 208
Sand-Grouse, 164, 168
Sand-Lark, 191
Sandpiper, 188, 205
Sandpiper, American Pectoral, 261
Sandpiper, Baird's, 261
Sandpiper, Bartram's, 261
Sandpiper, Bonaparte's, 261
Sandpiper, Broad-billed, 261
Sandpiper, Buff-breasted, 261
Sandpiper, Common, 206, 210
Sandpiper, Curlew-, 206
Sandpiper, Green, 211, 212
Sandpiper, Marsh-, 261
Sandpiper, Purple, 207
Sandpiper, Semipalmated, 261
Sandpiper, Siberian Pectoral, 261
Sandpiper, Solitary, 212, 261
Sandpiper, Spotted, 261
Sandpiper, Terek, 261
Sandpiper, Wood-, 211, 212
Sawbill, 162, 163
Saxicola indica, 257
Saxicola rubetra, 24
Saxicola rubicola, 24
Scarf, 134
Scart, 134
Scolopax rusticula, 200
Scooper, 198

Scoter, Common, 160, 161
Scoter, Surf-, 161, 260
Scoter, Velvet, 160
Sea-parrot, 241
Sea-pie, 197
Sea-snipe, 205
Sea-swallow, 219
Serinus serinus, 258
Shag, 133, 134
Shearwater, 253
Shearwater, Dusky, 253
Shearwater, Great, 253, 254
Shearwater, Levantine, 254, 261
Shearwater, Little Madeiran, 262
Shearwater, Manx, 254
Shearwater, Mediterranean Great, 262
Shearwater, Sooty, 253, 262
Sheldrake, 149
Sheldrake, Ruddy, 260
Shoe-horn, 198
Shoveler, 152
Shrike, 61
Shrike, Corsican Woodchat, 258
Shrike, Great Grey, 62
Shrike, Lesser Grey, 63
Shrike, Masked, 258
Shrike, Red-backed, 61, 63
Shrike, Southern Grey, 258
Shrike, Woodchat, 62
Siskin, 72
Sitta cæsia, 51
Sittidæ, 15, 51
Skua, 219, 232
Skua, Arctic, 233–235
Skua, Buffon's, 235
Skua, Great, 232–234
Skua, Long-tailed, 235
Skua, Pomatorhine, 235
Skua, Richardson's, 233, 235
Skylark, 95, 97
Skylark, Eastern, 259
Smew, 163
Snipe, 188, 202, 203
Snipe, Common, 202, 203
Snipe, Great, 203
Snipe, Jack, 203, 204
Snipe, Red-breasted, 261
Snipe, Solitary, 203
Solan Goose, 134
Somateria mollissima, 159
Somateria spectabilis, 260

Index

Sparrow, House-, 73
Sparrow, Tree-, 73
Spatula clypeata, 152
Spinus citrinella, 258
Spinus spinus, 72
Spoonbill, 136, 142, 143
Sprosser, 257
Squatarola squatarola, 193
Starling, 85, 86
Starling, Rose-coloured, 86
Steganopodes, 16, 132
Stercorariidæ, 17, 219, 232
Stercorarius longicaudus, 235
Stercorarius parasiticus, 233
Stercorarius pomatorhinus, 235
Sterna, 222
Sterna anglica, 261
Sterna caspia, 261
Sterna dougalli, 223
Sterna fuliginosa, 261
Sterna hirundo, 224
Sterna minuta, 225
Sterna paradisea, 224
Sterna sandvicensis, 222
Sterninæ, 17, 219
Stilt, 188
Stilt, Black-winged, 261
Stint, 205
Stint, American, 261
Stint, Little, 205, 206
Stint, Temminck's, 205
Stonechat, 23, 24, 28
Stonechat, Indian, 257
Stone-Curlew, 188, 189
Stork, 136, 184
Stork, Black, 142, 260
Stork, White, 141, 142
Stormcock, 19
Streptopelia orientalis, 260
Streptopelia turtur, 167
Striges, 16, 111
Strigidæ, 16, 112
Strix aluco, 114
Sturnidæ, 15, 85
Sturnus vulgaris, 85, 86
Subfamily, 15–17
Suborder, 17
Sula bassana, 134
Surnia ulula, 259
Surnia ulula caparoch, 259
Swallow, 66–69
Swallow, Red-rumped, 258

Swan, 143, 144, 147
Swan, Bewick's, 148
Swan, Mute, 148
Swan, Whooper, 144, 148
Swift, 67, 99, 100, 129
Swift, Alpine, 259
Swift, Needle-tailed, 259
Swift, White-bellied, 259
Sylvia atricapilla, 30
Sylvia communis, 29
Sylvia curruca, 30
Sylvia melanocephala, 257
Sylvia nisoria, 257
Sylvia orphea, 257
Sylvia simplex, 30
Sylvia subalpina, 257
Sylviinæ, 15, 18, 22, 29
Syrrhaptes paradoxus, 168

Tadorna casarca, 260
Tadorna tadorna, 149
Tammie Norie, 241
Teal, 144, 153, 155, 156
Teal, American Green-winged, 260
Teal, Blue-winged, 260
Teal, Summer-, 153
Terekia cinerea, 261
Tern, 219–225
Tern, Arctic, 224
Tern, Black, 220
Tern, Caspian, 222, 261
Tern, Common, 224
Tern, Gull-billed, 222, 261
Tern, Little, 225
Tern, Noddy, 221, 261
Tern, Roseate, 223
Tern, Sandwich, 222
Tern, Sooty, 221, 261
Tern, Whiskered, 222
Tern, White-winged Black, 222
Tetrao urogallus, 170
Tetraonidæ, 17, 69
Thalassidroma pelagica, 251
Thrush, Black-throated, 257
Thrush, Common, 19
Thrush, Continental Song-, 257
Thrush, Dusky, 257
Thrush, Mistle-, 19
Thrush, Rock-, 257
Thrush, White's, 257
Thrushes, 18, 39
Tichodroma muraria, 258

Tit (Titmouse), 43–46, 53
Tit, Bearded, 42
Tit, Blue, 46, 49, 55
Tit, Coal, 46, 47, 49
Tit, Continental Coal, 258
Tit, Crested, 49
Tit, Great, 46, 51
Tit, Long-tailed, 44
Tit, Marsh-, 48, 49
Tit, Northern Willow-, 258
Tit, White-headed Long-tailed, 258
Tit, Willow-, 48, 49
Titlark, 58, 59
Totanus, 211
Totanus flavipes, 261
Totanus fuscus, 214
Totanus glareola, 211
Totanus macularius, 261
Totanus melanoleucus, 261
Totanus nebularius, 214
Totanus ochropus, 211
Totanus solitarius, 261
Totanus stagnatilis, 261
Totanus totanus, 212
Tringa acuminata, 261
Tringa alpina, 204
Tringa bairdi, 261
Tringa canutus, 207, 208
Tringa ferruginea, 206
Tringa fuscicollis, 261
Tringa maculata, 261
Tringa maritima, 207
Tringa minuta, 205
Tringa minutilla, 261
Tringa temmincki, 205
Tringoides hypoleucus, 210
Troglodytes troglodytes, 52
Troglodytidæ, 15, 52
Tryngites subruficollis, 261
Tubinares, 17, 250
Turdidæ, 15, 18
Turdinæ, 15, 18
Turdus atrigularis, 257
Turdus aureus, 257
Turdus fuscatus, 257
Turdus iliacus, 20
Turdus merula, 20
Turdus musicus, 257
Turdus musicus clarkii, 19
Turdus pilaris, 22
Turdus torquatus, 20
Turdus torquatus alpestris, 257

Turdus viscivorus, 19
Turnix, 178
Turnix sylvatica, 260
Turnstone, 12, 195, 196, 207, 209
Turtle-Dove, 167
Turtle-Dove, Eastern, 260
Twite, 76
Tystie, 240

Upupa epops, 108
Upupidæ, 16, 108
Uria grylle, 240
Uria lomvia, 240
Uria troile, 238

Vanellus vanellus, 194
Vulture, 116
Vulture, Egyptian, 117, 260
Vulture, Griffon, 116, 117, 259
Vulturidæ, 16

Wagtail, 55, 56
Wagtail, Ashy-headed Yellow, 258
Wagtail, Black-headed, 258
Wagtail, Blue-headed, 58
Wagtail, Grey, 57, 58
Wagtail, Pied, 56, 58, 111
Wagtail, Sykes' Yellow, 258
Wagtail, Water-, 56, 58
Wagtail, Yellow, 36, 57, 58
Warbler, 29
Warbler, Aquatic, 258
Warbler, Arctic Willow-, 257
Warbler, Barred, 257
Warbler, Blyth's Reed-, 258
Warbler, Cetti's, 258
Warbler, Dartford, 31
Warbler, Dusky, 257
Warbler, Garden, 28, 30, 31
Warbler, Grasshopper-, 36
Warbler, Great Reed-, 258
Warbler, Greenish, 257
Warbler, Icterine, 257
Warbler, Marmora's, 32
Warbler, Marsh-, 34–36
Warbler, Melodious, 257
Warbler, Orphean, 257
Warbler, Pallas's Grasshopper-, 258
Warbler, Radde's Grasshopper-, 258
Warbler, Reed-, 34–38, 110
Warbler, Rufous, 258

Index

Warbler, Russian Willow-, 257
Warbler, Sardinian, 257
Warbler, Savi's, 38
Warbler, Sedge-, 34–36, 110
Warbler, Subalpine, 257
Warbler, Syrian, 258
Warbler, Temminck's Grasshopper-, 258
Warbler, Willow-, 32
Warbler, Yellow-browed, 257
Watching Birds, 9
Waterhen, 179, 182
Waxwing, 63, 64
Whaup, 216
Wheatear, 22, 28
Wheatear, Black, 257
Wheatear, Eastern Black-eared, 257
Wheatear, Eastern Desert-, 257
Wheatear, Greenland, 257
Wheatear, Isabelline, 257
Wheatear, Pied, 257
Wheatear, Western Black-eared, 257
Wheatear, Western Black-throated, 257
Wheatear, Western Desert-, 257
Whimbrel, 218

Whinchat, 24
Whitethroat, Common, 29, 30, 31
Whitethroat, Lesser, 30
Whooper, see Swan
Wigeon, 155, 156
Wigeon, American, 260
Willock, 238
Windhover, 130
Woodcock, 188, 200–202
Woodpecker, Great Spotted, 105
Woodpecker, Green, 105
Woodpecker, Lesser Spotted, 105
Woodpecker tribe, 99, 102
Woodpigeon, 164, 166, 171
Wren, 39, 52
"Wren," Fire-crested, 50, 51
"Wren," Gold-crested, 50
"Wren," Willow-, 32–34
"Wren," Wood-, 33
Wryneck, 99, 102, 103

Xema sabinii, 225

Yaffle, 105
Yellowhammer, 81, 82
Yellowshank, 261
Yellowshank, Greater, 261

www.ingramcontent.com/pod-product-compliance
Ingram Content Group UK Ltd.
Pitfield, Milton Keynes, MK11 3LW, UK
UKHW040657180125
453697UK00010B/242